THE TSWANA

THE TSWANA

BY

I. SCHAPERA

WITH A SUPPLEMENTARY CHAPTER BY

JOHN L. COMAROFF

AND A SUPPLEMENTARY BIBLIOGRAPHY BY

ADAM KUPER

KPI

IN ASSOCIATION WITH THE
INTERNATIONAL AFRICAN INSTITUTE

*First published in 1953, as part of the
Ethnographic Survey of Africa
edited by Daryll Forde
This edition published in 1984 by
KPI Limited*

*Routledge & Kegan Paul PLC
14 Leicester Square, London WC2H 7PH, England*

*Routledge & Kegan Paul
9 Park Street, Boston, Mass. 02108, USA*

*Routledge & Kegan Paul
464 St Kilda Road, Melbourne,
Victoria 3004, Australia, and*

*Routledge & Kegan Paul
Broadway House, Newtown Road,
Henley-on-Thames, Oxon RG9 1EN, England*

*Printed in Great Britain by
Hartnoll Print, Bodmin, Cornwall*

© *International African Institute 1953, 1968, 1976, 1984*

*No part of this book may be reproduced in
any form without permission from the publisher,
except for the quotation of brief passages
in criticism*

ISBN 0-7103-0096-4

FOREWORD

THE International African Institute has, since 1945, been engaged on the preparation and publication of an Ethnographic Survey of Africa, the purpose of which is to convey in a brief and readily comprehensible form a summary of available information concerning the different peoples of Africa with respect to location, natural environment, economy and crafts, social structure, political organization, religious beliefs and cults. While available published material has provided the basis for the Survey, a mass of unpublished documents, reports and records in Government files and in the archives of missionary societies, as well as field notes and special communications by anthropologists and others, have been generously made available, and these have been supplemented by personal correspondence and consultation. The Survey is being published in a number of separate volumes, each of which is concerned with one people or a group of related peoples, and contains a comprehensive bibliography and a specially drawn map.

A committee of the Institute was set up under the Chairmanship of Professor Radcliffe-Brown, and the Director of the Institute has undertaken the organization and editing of the Survey. The generous collaboration of a number of research institutions and administrative officers in Europe and in the African territories was secured, as well as the services of senior anthropologists who have been good enough to supervise and amplify the drafts.

The work of the Survey was initiated with the aid of a grant from the British Colonial Development and Welfare Funds, on the recommendation of the Social Science Research Council, to be applied mainly though not exclusively to work relating to British territories. A further grant from the Sudan Government has assisted in the preparation and publication of sections dealing with that territory.

The Ministère de le France d'Outre-mer and the Institut Français d'Afrique Noire were good enough to express their interest in the project and through their good offices grants have been received from the Governments of French West Africa and the French Cameroons for the preparation and publication of sections relating to these areas. These sections have been prepared by French ethnologists with the support and advice of Professor M. Griaule of the Sorbonne and Professor Th. Monod, Director of I.F.A.N.

The collaboration of the Belgian authorities in this project was first secured by the good offices of the late Professor de Jonghe, who enlisted the interest of the Commission d'Ethnologie of the Institut Royal Colonial Belge. The collaboration of the Institut pour la Recherche Scientifique en Afrique Centrale has also been readily accorded. Work relating to Belgian territories is being carried out under the direction of Professor Olbrechts at the Centre de Documentation of the Musée du Congo Belge, Tervuren, where Mlle. Boone and members of her staff are engaged on the assembly and classification of the vast mass of material relating to African peoples in the Belgian Congo and Ruanda-Urundi. They work in close collaboration with ethnologists in the field to whom draft manuscripts are submitted for checking.

Since the unequal value and unsystematic nature of existing material was one of the reasons for undertaking the Survey, it is obvious that these studies cannot claim to be complete or definitive; it is hoped, however, that they will present a clear account of our existing knowledge and indicate where information is lacking and further research is needed; a list of those sections which have already appeared will be found on p. 79 of this volume.

The International African Institute desires to express its very grateful thanks to those official bodies whose generous financial assistance has made the carrying out of this project possible and to the many scholars, directors of research organizations,

FOREWORD

administrative officers, missionaries, and others who have collaborated in the work and, by granting facilities to our research workers and by correcting and supervising their manuscripts, have contributed so largely to whatever merit the sections may possess.

In connection with the preparation of this volume the author and the Institute are grateful for assistance and information provided by Professor A. N. Tucker, Dr. V. G. J. Sheddick, the Bechuanaland Protectorate Administration, the Director of Census and Statistics (Union of South Africa), and Mr. A. T. Carey.

DARYLL FORDE,
Director,
International African Institute.

CONTENTS

	PAGE
GROUPINGS, DEMOGRAPHY, AND HISTORY	9
Nomenclature	9
Location and Groupings	9
Demography	11
History	14
LANGUAGE AND LITERATURE	17
ECONOMY	19
Physical Environment	19
Land tenure	20
Production of food	21
Agriculture	21
Animal husbandry	22
Hunting	24
Food and narcotics	25
Arts and crafts	25
Clothing and ornaments	25
Dwellings and household goods	26
Tribute and taxation	26
Organization of labour	27
Household production	27
Specialized crafts	27
Labour assistance	28
Retainers and serfs	28
Regimental labour	28
Trade and exchange	29
Wage-labour	30
Local employment	30
Labour migration	30
Tribal finances	31
SOCIAL ORGANIZATION	34
The Tribe	34
Totemism	35
Territorial organization	35
Social status	36
Rank and social classes	36
Sex and age differentiation	37
Age-sets	38
Domestic groupings	39
The household	39
The family-group	40
Marriage	41
Divorce	42
Succession and inheritance	42
Kinship	43
Wards and sections	46
Modern groupings	47

CONTENTS—*continued*

	PAGE
GOVERNMENT AND LAW	49
European control	49
Central government	49
Native councils	49
Local administration	50
Tribal administration	51
Chiefship	51
The chief's advisers and councils	52
Local government	53
Law and judicial procedure	55
Tribal law	55
Courts and their jurisdiction	55
Conduct of trials	56
RELIGION AND MAGIC	58
Religion	58
Christianity	58
Traditional cults	59
Modern survivals	61
Magicians and magic	61
Magicians	61
Morphology of magic	63
Divination	64
Sorcery	65
TSWANA TRANSFORMATIONS, 1953–1975	67
The national context	67
South Africa	67
Botswana	69
Tswana transformations	70
Government and politics	71
Social organization	73
Conclusion	75
SELECT BIBLIOGRAPHY	77
SUPPLEMENTARY BIBLIOGRAPHY 1953–1967	85
SUPPLEMENTARY BIBLIOGRAPHY 1968–1975	87
Index	90

GROUPINGS, DEMOGRAPHY, AND HISTORY

Nomenclature[1]

The Tswana (*BaTswana*, sing. *MoTswana*) are one of the three major divisions into which ethnologists and linguists usually classify the Sotho group of Bantu-speaking peoples of central South Africa. The two others are distinguished as " Southern Sotho " (of Basutoland) and " Northern, Eastern, or Transvaal, Sotho " (of the northern and eastern Transvaal); the Tswana, correspondingly, are sometimes also termed " Western Sotho."[2] All members of the division nowadays accept and use BaTswana as a common name, although some eastern tribes (e.g., Kgatla) occasionally also call themselves BaSotho.

The origin of the name BaTswana is uncertain. It has been variously interpreted as meaning: (*a*) " the little offshoots " (from *-tswa*, " to go out, to come from "), (*b*) "the separatists, or seceders" (from *-tswaana*, "to separate from one another"), and (*c*) " those who are alike " (from *-tshwaana*, " to be alike ").[3] None of these derivations is generally accepted; van Warmelo curtly dismisses them as " fruitless speculation."[4] Early European writers recorded the name in many different forms (e.g., Beetjuana, Bichwana, Booshuana, Bootchuana, and Buchuana), but it soon became standardized as Bechuana (whence Bechuanaland, the country inhabited by the people); the variant BeChwana is occasionally also used.

The southern tribes, especially the Tlhaping, were called *Birina* or *Birikwa* (" the goat people ") by the Hottentots, and the Tawana in the north-west are called *BaRwa* by the Koba, Mbukushu, and other Western Bantu; but to most of their neighbours the Tswana are usually known by some local variant of that name.

Location and Groupings

Today the Tswana are found all over the Bechuanaland Protectorate, but especially in the eastern and north-western portions; in the western and central districts of the Transvaal (especially Marico, Rustenburg, Pretoria, Ventersdorp, and Lichtenburg); and in the northern districts of the Cape Province, collectively known as British Bechuanaland (especially Mafeking, Vryburg, Kuruman, and Taungs). There are also offshoots in the Orange Free State (Thaba Nchu district), Southern Rhodesia (Plumtree district), and South West Africa (Gobabis district), but the last two are very small and politically insignificant. Their principal neighbours, apart from Europeans, are Kalanga (Kalaka) and other Shona in the north-east, Central Sotho and Ndebele (Tebele) in the Transvaal, Kgalagadi and Bushmen (Sarwa) in the west, and Herero (Tamma), Yeei (Koba), and Gova (Mbukushu), in the north-west;[5] all, except the Bushmen, are also Bantu-speaking peoples. Many of them, especially in the Protectorate, inhabit the same areas as the Tswana and are subject to Tswana chiefs; the population of a tribe may therefore consist of both Tswana and peoples of alien origin.[6]

The Tswana themselves seem on the whole to be sufficiently homogeneous to be classed as a single group in relation to the other peoples of South Africa. Local variations occur, both in dialect and in social structure and other aspects of culture, but much more needs to be learned, especially about the tribes in the Union, before a definitive classification into sub-groups can be attempted. On the available

---

[1] In this study the orthography used is that officially recognized by the European Administrations.
[2] Lestrade, 1929, p. 8 ; van Warmelo, 1935, pp. 96 f. The Kgalagadi, hitherto regarded as impoverished Tswana, probably constitute a fourth division of the Sotho ; see below, p. 14.
[3] Stow, 1905, p. 408 ; Brown, 1926, pp. 25 f.
[4] 1935, p. 103.
[5] The names in parentheses are those commonly used by the Tswana.
[6] See below, p. 34.

evidence, two sub-groups may be distinguished: Western and Eastern.[7] The differences between them are partly geographical, partly cultural and historical; details will be given below where relevant. Each sub-group is composed of several " clusters " and many different " tribes "; a tribe is a politically independent unit, with its own chief and territory, and a cluster consists of several tribes, which were at one time united under the rule of a single chief and which often still bear a common name. The principal clusters and tribes, and their present distribution, are as follows:[8]

WESTERN TSWANA:
Tlhaping
    Phuduhutswana   C. Taungs, Vryburg, Barkly West
    Maidi   C. Taungs
Rolong
    Rratlou   T. Lichtenburg ; C. Mafeking, Vryburg
    Tshidi   C. Mafeking ; P. Lobatsi
    Seleka   O. Thaba Nchu ; P. Francistown
    Rrapulana   T. Lichtenburg ; C. Mafeking
    Tloung   T. Lichtenburg
    Kaa   P. Mochudi, Ngwato, Kweneng
    Kubung   T. Ventersdorp
Hurutshe
    Manyana   P. Ngwaketsi ; T. Marico
    Mokhubidu   P. Kweneng ; T. Marico
    Gôpane   T. Marico
    Moilwa   T. Marico
    Khurutshe   P. Francistown
    Tlharo   C. Mafeking, Vryburg, Kuruman
Nogeng   T. Lichtenburg
Kwena   P. Kweneng
    Ngwaketse   P. Ngwaketsi
    Ngwato   P. Ngwato
    Tawana   P. Ngamiland

EASTERN TSWANA:
(a) Kwena
    Fokeng   T. Rustenburg, Ventersdorp
    Mogôpa   T. Pretoria, Rustenburg, Ventersdorp
    Mmanamêla   T. Rustenburg
    Modimosana   T. Rustenburg
    Mmatau   T. Rustenburg
    Matlaku   T. Rustenburg
    Phalane   T. Rustenburg, Marico
Tlhalerwa   T. Rustenburg
Phiring   T. Rustenburg
Taung   T. Rustenburg
Kgatla
    Mosêtlha   T. Pretoria
    Kgafêla   P. Mochudi ; T. Rustenburg
    Mmanaana   P. Kweneng, Ngwaketsi
    Mmakau   T. Pretoria
    Motšha   T. Pretoria
    Seabe   T. Pretoria
Bididi   T. Waterberg
Tlôkwa
    Thethe   P. Gaberones ; T. Rustenburg
    Motsatsie   T. Rustenburg
(b) Malete   P. Gaberones ; T. Marico
Tlhako   T. Rustenburg
Seleka   T. Potgietersrust
Pô   T. Rustenburg
Hwaduba   T. Pretoria

[7] van Warmelo, 1935, pp. 103, 106.
[8] van Warmelo, 1935, pp. 103–8 ; Schapera, 1952 (b), p. 30 and *passim*. The localities named are the magisterial districts within which tribal territories are situated. The initial letter preceding the name of a district indicates the country or province : P = Bechuanaland Protectorate, T = Transvaal, C = Cape Province, and O = Orange Free State.

The people listed above as "Eastern Tswana, (b)" are in fact Transvaal Ndebele (i.e., Nguni) by origin, but they have so completely adopted the language and culture of the Tswana that they are nowadays usually classed as members of that group. Some of the others (e.g., Kaa and Mmanaana-Kgatla) were at one time independent tribes, but are now subject communities; on the other hand, the Tlhaping, Rolong, Hurutshe, Kwena, and several other clusters, all include some very small tribes not shown on the list. In addition, minor offshoots from virtually every tribe are now also found as subject communities in one or more of the others (e.g., there are communities of Kwena in all the tribes of the Protectorate).[9]

DEMOGRAPHY

According to the 1946 census, there are 579,800 Tswana in the Union of South Africa (Transvaal 365,000, Cape Province 168,400, Orange Free State 43,800, and Natal 2,600); in the Protectorate, whose total Native population is 292,800, they number 270,500 (of whom 103,500 are peoples of alien origin living under the rule of Tswana chiefs).[10] If we add 1,700 (probably a generous estimate) for Tswana in Southern Rhodesia and South-West Africa, the total population of the group (including alien subjects) may be estimated at approximately 852,000, chiefly distributed as follows: Transvaal, 43%, Protectorate 32%, Cape Province 20%, and Orange Free State 5%.

Owing to European encroachments, most Tswana tribes are now confined to areas much smaller than they once occupied and, especially in the Union, many people live on land owned by Europeans, to whom they either render service or pay rent in cash. Tribally-held lands embrace an area of about 105,000 sq. miles in the Protectorate, 6,000 sq. miles in British Bechuanaland (Cape Province), and 3,600 sq. miles in the Transvaal. The following table shows, for each of the districts where Tswana constitute all or the great bulk of the African population,[11] (a) the total area (in sq. miles), (b) the area owned by or reserved for Africans, (c) the percentage proportion of the African area, (d) the African population of that area, and (e) the density of the population (per sq. mile).[12]

District	Total area	Native area	%	Population	Density
Protectorate [13]					
*Ngamiland	34,500	34,500	100.0	38,700	1.1
*Ngwato	42,000	42,000	100.0	101,000	2.4
*Kweneng	15,000	15,000	100.0	39,800	2.7
*Ngwaketsi	9,000	9,000	100.0	38,600	4.3
*Mochudi	3,600	3,600	100.0	20,100	5.6
Gaberones	495	245	49.5	11,700	47.8
Lobatsi	665	430	64.7	4,700	10.9
	105,260	104,775	99.5	254,600	2.4

[9] Schapera, 1952 (b), pp. 124 f.
[10] Figures for the Union supplied by the Director of Census and Statistics; those for the Protectorate compiled from the preliminary census returns (December, 1946). All figures are to the nearest 100.
[11] In other districts inhabited by Tswana, e.g., Pretoria, there are also many Africans of other groups; the census returns do not distinguish the various groups, so that it is impossible to give data for Tswana alone. Such districts have therefore been omitted from the Table.
[12] Data for the Union derived from *Population Census*, 1946, Vol. I (U.G.51, 1949), Table 10, and Report No. 9 of the Social and Economic Planning Council (U.G.32, 1946), Table I; for the Protectorate, from 1946 Census Report, Table II (c), and Schapera, 1943 (a), pp. 13 f., 57. Population figures are given to the nearest 100, and area figures to the nearest 5.
[13] The five districts marked with an asterisk, each consisting of a single tribal Reserve, have not yet been surveyed; the areas given are the official estimates.

District	Total area	Native area	%	Population	Density
Transvaal:					
Rustenburg	9,020	1,845	20.5	56,600	30.7
Marico	3,675	760	20.7	22,600	29.7
Lichtenburg	4,295	530	12.3	14,800	27.9
	16,990	3,135	18.5	94,000	30.0
Cape Province:					
Mafeking	4,265	1,215	28.5	31,300	25.7
Vryburg	16,195	2,800	17.3	18,800	6.7
Kuruman	13,920	920	6.6	13,600	14.8
Taungs	1,440	675	46.9	19,800	29.3
	35,820	5,610	15.7	83,500	14.9
O.F.S.:					
Thaba Nchu	1,185	240	20.3	14,800	61.6

The extent to which Europeans and other non-Africans have penetrated into "Bechuanaland" is shown by the following table, which gives (a) the number of Africans living on rural land owned by Europeans in each of the districts listed above, (b) the total African population of each district,[14] (c) the European population, (d) the number of Asiatics and Coloured, and (e) the total population (all ethnic groups).[15]

District	Africans on Eur. land	Total Africans	Europeans	Others	Total
Protectorate:					
Ngamiland	—	38,720	130	10	38,860
Ngwato	—	100,990	510	150	101,650
Kweneng	—	39,830	120	180	40,130
Ngwaketsi	—	38,560	50	190	38,800
Mochudi	—	20,110	70	30	20,210
Gaberones	300	12,030	190	90	12,310
Lobatsi	3,200	7,900	410	50	8,360
	3,500	258,140	1,480	700	260,320
Transvaal:					
Rustenburg	35,690	108,290	28,550	1,420	138,260
Marico	18,530	45,730	10,190	810	56,730
Lichtenburg	36,400	63,630	17,370	950	81,950
	90,620	217,650	56,110	3,180	276,940
Cape Province:					
Mafeking	7,190	41,050	6,540	950	48,540
Vryburg	19,370	41,930	10,160	2,030	54,120
Kuruman	9,370	28,300	6,740	2,490	37,530
Taungs	4,540	26,400	5,640	500	32,540
	40,470	137,680	29,080	5,970	172,730
O.F.S.:					
Thaba Nchu	13,720	29,670	2,240	1,010	32,920

[14] The figure includes people living in urban areas, mine compounds, etc., of whom a large proportion are almost certainly not Tswana.
[15] *Population Census of the Union,* 1946, Vol. I, Tables 7 and 10; *B.P. Census,* 1946, **Table II** and preliminary returns (which contain some information not published in the final report). All population figures are to the nearest 10.

Comparison of the 1946 census figures with those for 1936 shows that the population has apparently increased. The regional totals for the two years, and the rate of increase, are as follows:[16]

	Population		Increase	
	1936	1946	Persons	%
Protectorate	223,867	258,141	34,274	15.3
Transvaal	165,440	200,808	35,368	21.4
Cape Province	110,658	129,576	18,918	17.1
O.F.S.	22,618	29,589	6,971	30.8
	522,583	618,114	95,531	18.3

During the same period, the percentage increase of the African population in the Union as a whole was 18.7 (Transvaal 27.7, Cape 14.3, and O.F.S. 19.9), and in the Protectorate as a whole 12.6.[17] However, the 1936 census returns, especially for the Protectorate, are so unreliable that the figures given above must not be taken literally. No vital statistics of the Native population are kept, in either the Union or the Protectorate, and the available data are insufficient to permit of any confident statement about the rate at which it is in fact increasing.[18]

The 1946 census returns show also that, in the Union, females greatly outnumber males inside the tribal areas. The district figures are as follows:

District	Males	Females	Masculinity [19]
Rustenburg	25,250	31,350	81
Marico	9,700	12,900	75
Lichtenburg	7,010	7,770	90
	41,960	52,020	81
Mafeking	14,130	17,130	83
Vryburg	8,490	10,280	82
Kuruman	5,990	7,630	79
Taungs	9,120	10,680	85
	37,730	45,720	83
Thaba Nchu	6,840	7,990	86
TOTAL	86,530	105,730	82

The preponderance of females is, almost certainly, due very largely to labour migration of the males. This is suggested by the figures for the Protectorate, where the census returns give not only the number of people at home on the day of the census, but also the number absent in the Union and elsewhere.[20] As will be seen from the following table, the sexes are more evenly balanced if we consider the total

---

[16] The regional totals are for *all rural* Africans in the districts listed above. The data for 1936 do not permit separate treatment of the people living in tribal areas only.
[17] *Population Census of the Union*, 1946, Vol. I, Tables 4 and 7; *B.P. Census*, 1946, p. (iv) and Table II (a).
[18] Cf. the discussion in Schapera, 1947 (a), pp. 9–24; Kuczynski, 1949, pp. 72–81; and *B.P. Census Report*, 1946, pp. (vii) f.
[19] Masculinity = number of males for every 100 females.
[20] *Census Report*, 1946, Table II (h).

population and not merely the people at home. (The causes and extent of labour migration are discussed below, in the section on " Economy.")[21]

District	At home			Total population		
	M.	F.	Masc.	M.	F.	Masc.
Ngamiland	18,620	19,760	92	18,940	19,780	96
Ngwato	47,380	50,040	95	50,530	50,460	100
Kweneng	17,860	19,920	90	19,750	20,070	98
Ngwaketsi	17,850	19,270	93	19,070	19,490	98
Mochudi	8,480	9,610	88	9,880	10,230	97
Gaberones	4,170	5,480	76	5,720	6,020	95
Lobatsi	1,920	2,260	85	2,310	2,400	96
	116,280	126,340	92	126,200	128,450	98

The age distribution of the population in the Protectorate shows that the proportions of the sexes vary from one age group to another.[22] " What significance could be attached to these tendencies were mortality rates available, it is difficult to say, but the figures do at least suggest some consistently greater mortality of one sex over the other at certain ages."[23]

Age Group	Numbers			Percentage	
	M.	F.	Persons	M.	F.
Under 5	13,310	13,380	26,690	49.9	50.1
5-15	36,870	35,760	72,630	50.8	49.2
16-50	60,720	64,470	125,190	48.5	51.5
Over 50	16,290	15,880	32,170	50.6	49.4
Unspecified	750	720	1,470	—	—
	127,940	130,210	258,150	49.6	50.4

History

Nothing definite is known about the origins of the Sotho peoples to whom the Tswana belong. The conventional view is that they separated from the main body of Bantu-speaking peoples somewhere in the vicinity of the Great Lakes of East Africa, and that they entered South Africa, probably through the western portions of Southern Rhodesia, in three series of migrations.[24] The first is represented today by the people collectively known as Kgalagadi, who settled in the eastern, more fertile, parts of Bechuanaland, where they intermingled freely with the pre-existing Sarwa (Bushman) population. They were at one time commonly described as culturally-degenerate Tswana, but recent research has shown that their language is sufficiently distinct to be classed as a separate (and fourth) member of the Sotho group.[25] Following upon them came the ancestors of the modern Rolong and Tlhaping, who settled along the upper reaches of the Molopo River, from which they gradually spread south and west. They absorbed some of their Sarwa and Kgalagadi predecessors, most of whom, however, retreated before them into the arid

[21] See p. 30.
[22] The figures (extracted from the 1946 *Census Report*, Table IV (a)) are for the total Native population of the districts named above ; no data are available for the tribal areas only. The Union figures have not yet been published.
[23] *B.P. Census*, 1946, Report (by C. W. Cousins), p. (v).
[24] Stow, 1905, chapters xxi f. ; Lestrade, 1929, pp. 10 f.
[25] van der Merwe and Schapera, 1943, p. 3.

zones of the Kalahari Desert. The third, and greatest, migration brought the ancestors of all the other Sotho tribes. They settled, as a united body, in the southwestern portions of the modern Transvaal, and then broke up rapidly into separate clusters, the most important of which were the Hurutshe, Kwena, and Kgatla.

Even if the traditions are reasonably accurate, it is impossible to date the migrations they record. All that can be said with some confidence is that the Tswana were already in the eastern half of their present habitat by about A.D. 1600. During the next two centuries, each of the existing clusters became increasingly subdivided. It was a constantly recurring feature in Tswana history for part of a tribe to secede under a discontented member of the ruling family and move away to a new locality. There it would set up as an independent tribe under the chieftainship of its leader, by whose name it generally came to be known. The Rolong, for instance, broke up into the Tlhaping, Maidi, Kaa, and various other tribes still usually called Rolong but more specifically identified by the names of their founders (Rratlou, Tshidi, Seleka, Rrapulana, Modibowa, Mosadi, etc.); the Hurutshe, said to have been the leaders of the final migration from the north, broke up into the modern Hurutshe tribes (Manyana, Mokhubidu, Moilwa, Gôpane, etc.), Tlharo, Khurutshe, and others; the Kgatla, an early offshoot from the Hurutshe, broke up into the modern Kgatla tribes (Mosêtlha, Kgafêla, Mmanaana, Mmakau, Motšha, etc.), Tlôkwa (Thethe and Motsatsie), and Pedi (now classed as Central Sotho); and the Kwena, another early offshoot from the Hurutshe, broke up into the Fokeng, Mogôpa, Modimosana, Phalane, etc. A section of the Mogôpa subsequently seceded and moved westwards (c. 1720); these people, now known as the Kwena of Sechele, were the first important Tswana tribe to settle in the present Protectorate. The Ngwaketse and Ngwato broke away from them not long afterwards, and the Tawana then broke away from the Ngwato (c. 1795). These are only some of the new tribes that came into being before the end of the 18th century.[26]

The process of fission indicated above has continued ever since; almost every one of the existing tribes has broken apart, sometimes more than once, because of internal dispute or some other factor. But relatively few of the groups seceding within more recent times succeeded in becoming independent. One reason is that from about 1810 to 1840 there prevailed among the Tswana a period of chaos, due initially in some tribes to civil wars, but mainly to the successive onslaughts of invaders from the east, notably the MmaNtatisi (1822–3), Sebetwane's Kololo (1823–8), and Moselekatse's Tebele (1825–37). During this time some of the Tswana tribes were forced to flee from their homes, to which they did not return until the danger was past; others, less fortunate, were irretrievably broken up into scattered groups that managed to survive only by attaching themselves permanently to other tribes.

Another factor inhibiting the creation of new tribes was the gradual extension of European control, which made it more and more difficult for seceding groups to find unoccupied or unclaimed land where they could live in independence. Owing to Government intervention, moreover, civil disputes were usually settled within the tribe; and where they did result in secession, the groups breaking away either had to become subject communities elsewhere or return to their former allegiance.[27]

The impact of Western civilization coincided with the beginning of the 19th century. In 1801 the Tlhaping, the southernmost tribe, were reached by a small party of explorers from the Cape, and in 1816 the first Christian mission to the Tswana was established among them. During the next 30 years most of the other tribes, except those in the far north, were visited sporadically by traders, hunters, and explorers, and mission stations were also established among the Rolong (1822), Hurutshe (1836), Mmanaana-Kgatla (1843), and Kwena (1846). Livingstone's dis-

[26] Schapera, 1952 (b), pp. 8–11, where references are also given to the earlier sources on tribal history.
[27] Schapera, 1952 (b), pp. 11–19.

covery of Lake Ngami (1849) opened up the road to the north, with its great wealth of ivory and other hunting spoils, and the number of European visitors rapidly increased.

Meanwhile, in 1837, the Boer Voortrekkers from the Cape had settled in the Transvaal, after defeating and expelling the Tebele. By 1852, when their independence was formally recognized by the British authorities, there were already some 20,000 of them in the country, organized into several small states which subsequently amalgamated to form the South African Republic. They established small townships and farmed the surrounding land; they also claimed all the local Natives as their subjects, and exacted forced labour from them. Their policy ultimately caused several tribes to move across the ill-defined border into the present Protectorate (1852–69). Among these were the Mmanaana-Kgatla, Malete, Tlôkwa, Kgafêla-Kgatla, and two groups of Hurutshe. Some afterwards returned to the Transvaal, but those named above have been in the Protectorate ever since.

The Boers also resented the increasing contacts of the western tribes with Englishmen, especially missionaries, whom they accused of hostile propaganda and of supplying firearms to the chiefs. They accordingly tried on several occasions to extend their boundary farther west, especially after 1868, when diamonds were discovered at Kimberley. These attempts led to armed conflict with such tribes as the Kwena, Rolong, and Tlhaping. The outcome was that in 1884 the British ultimately responded to Native appeals, and proclaimed a Protectorate over the country south of the Molopo and west of the Republic. Friction with the Boers persisted, and in March, 1885, the Protectorate was extended to include the tribes farther north. In September the southern half of the Protectorate became a Crown Colony, known as British Bechuanaland, which 10 years later was annexed to the Cape. It was intended at the same time (1895) to transfer the administration of the Protectorate (north of the Molopo) to the British South Africa Company, organized by Cecil Rhodes in 1889. The chiefs of the Ngwato, Kwena, and Ngwaketse went to England to protest, and it was ultimately agreed that, in return for certain land and other concessions on their part, their tribes would remain directly under the Crown.[28]

By the end of the 19th century, the whole territory of the Tswana had been partitioned among the Cape Colony in the south, Great Britain in the north, and the South African Republic in the east. Native locations or Reserves were set aside for them in the Transvaal (1884 onwards), British Bechuanaland (1889), and the Protectorate (1899 onwards), but much of the land formerly held by the tribes, especially in the Transvaal, was already owned and occupied by Europeans. The powers of the chiefs had everywhere been curtailed, and their people made to pay annual taxes to the European Governments. There were also mission stations, schools, and resident traders in every important tribe; European clothing and utensils had been widely adopted; and men from all over the territory were going out periodically to work on the mines and in other European areas. These influences were all intensified during the present century, especially after the Cape Colony and the Transvaal were incorporated into the Union of South Africa (1910). Both here and in the Protectorate the system of tribal administration was also reformed, economic and educational developments were promoted, and health services introduced. As a result, the tribal life of the Tswana now differs markedly, in some respects, from the pattern described by the first European visitors 150 years ago.

---

[28] The extension of European control over the Tswana tribes is described in most histories of South Africa. Reference may be made especially to : J. A. I. Agar-Hamilton, *The Road to the North* (1937) ; H. M. Hole, *The Passing of the Black Kings* (1932) ; J. Mackenzie, *Austral Africa* (1887) ; A. P. Newton and E. A. Benians (eds.), *Cambridge History of the British Empire*, Vol. viii : *South Africa* (1936) ; A. Sillery, *The Bechuanaland Protectorate* (1952) ; and E. A. Walker, *A History of South Africa*, revised edition (1940).

LANGUAGE AND LITERATURE

The Sotho group of languages, to which Tswana belongs, shares with other Bantu languages of South Africa such distinctive characteristics as a highly complex conjugational system for verbs, and the use of suffixes (instead of prefixes, as elsewhere in Bantu) to indicate the locative, diminutive, and augmentative, forms of nouns. It differs from them in having such features as a nine-vowel system, aspirated consonants instead of nasal compounds (e.g., *motho,* person, Nguni *umuntu*), no prefixes for nouns, the use of the immutable formative *ke* for the copula (expressed in Nguni by inflexion), and the use of " demonstrative pronouns in the indirect relative clause, whereas Nguni uses a relative concord."[1]

Tswana itself differs from other Sotho languages in various details of phonetics, grammar, and vocabulary; thus, unlike Southern Sotho, it has no clicks, and uses the velar fricative instead of the aspirate (e.g., Tswana *bogadi,* S. Sotho *bohadi*), and, unlike Northern Sotho, it has the relative ending *-ng* instead of *-xo* (e.g., Tswana *dirang,* N. Sotho *diraxo*). Dialectal differences within Tswana include the occurrence of Meinhof's noun class 11 (prefix *lo-*) in the south (e.g., Rolong), but not in the north and east (where it is merged with class 5, prefix *le-*); the use of *š* in the south for the *s* of other dialects (e.g., Rolong *tšwa,* " to come out," elsewhere *tswa*), and of *t* in the north for the *tl* of other dialects (e.g., Ngwato *tou,* elephant, elsewhere *tlou*); the use in the east (e.g., Kgatla) of the conjunctive *ge,* " if," for the *ha* or *fa* of other dialects; and the possessive concords *ja-* (class 5) and *jwa-* (class 14) in the south, and occasionally in the north, for the *la-* and *ba-* of the east. As already mentioned, the number and distribution of dialects has not yet been determined, although it is evident that at least three can be distinguished.[2]

In the northern parts of the Protectorate many Tswana also speak the languages of subject communities, e.g., Sarwa (Bushman) and Kalaka, but Tswana itself is the only vernacular officially recognized in the Territory. Throughout " Bechuanaland," moreover, both English and Afrikaans are widely understood and spoken, and many loan-words have been incorporated into Tswana, especially from Afrikaans. Many Tswana, in addition, have learned to read and write, and letter-writing is nowadays an important means of communication, especially between men working abroad and their relatives and friends at home.[3] The first school was started by Robert Moffat among the Tlhaping (*c.* 1825), and today there are either missionary or Government schools in virtually every tribal area. In the Protectorate, 33,190 (13.3%) of the people in Tswana districts were at school in 1946 or had previously attended; of these, 13,610 were males (10.8% of the male population) and 19,580 females (15.3% of the female population). Many more had become literate without ever going to school. The total figures, for both categories, are given in the table (p. 18).[4]

The early missionaries not only reduced the language to writing, but also published translations of the Bible and other religious works. Since then much has been written and published in Tswana. Apart from devotional and religious works, school readers, etc., this literature consists largely in records of folk-lore (myths, legends, fables, proverbs, riddles, songs, praise-poems, etc.); within recent years it has also come to include tribal histories, descriptions of custom, translations of Shakespeare, and original fiction and verse. Most of the authors are Europeans, but, following the lead of S. T. Plaatje, Tswana themselves have begun to publish,

---

[1] Doke, 1937, pp. 314–9; cf. Guthrie, 1948, pp. 66–70.
[2] Dialectal differences are illustrated in the texts published by Schapera (ed.), 1938; cf. van der Merwe and Schapera, 1943, and the grammatical notes in Lestrade, 1944.
[3] Cf. Schapera, 1933 (b), pp. 20–8.
[4] Compiled from *B.P. Census Report,* 1946, Table IV. Comparable figures for the Union have not yet been published.

	Males	Females	Persons
Vernacular:			
Read and write ..	23,250	23,980	47,230
Read only ..	1,830	2,350	4,180
	25,080	26,330	51,410
Percentage ..	19.9	20.5	20.2
English:			
Read and write ..	7,790	9,950	17,740
Read only ..	2,450	3,050	5,500
	10,240	13,000	23,240
Percentage ..	8.1	10.1	9.1

notably M. Kgasi, S. S. Mafonyane, D. P. Moloto, D. M. Ramoshoana, L. D. Rraditladi, and M. O. M. Seboni. There have also been several newspapers, e.g., *Molekudi ua Bechuana* (monthly, 1856-7), *Mokaeri oa Becuana* (monthly, 1857-9), *Mahoko a Becwana* (monthly, 1883-98), and *Mosupa-Tsela* (monthly, since 1913), all published by missionary societies, *Koranta ea Becoana* (weekly, 1901-8) and *Tsala ea Batho* (weekly, c. 1910-16), both published by Tswana, and, since 1947, *Naledi ya BaTswana*, a flourishing commercially-owned weekly especially valuable for its regular columns of local news.[5]

---

[5] For annotated bibliographies of Tswana literature, cf. Doke, 1933, pp. 77-85; Lestrade, 1944, pp. 22-7; Letele, 1944, pp. 161-71.

ECONOMY

Physical Environment[1]

The Tswana inhabit a long, relatively narrow, tract of land running south-east from the Okavango River to the upper reaches of the Limpopo, and then south-west to the Kuruman area just north of the Orange River. It extends over three drainage systems: the Okavango-Makarikari swamp and salt-pans in the north, the Upper Limpopo (Crocodile) and Marico Rivers in the east, and the Molopo-Kuruman area in the south. (a) The Okavango area, where the Tawana live, is a very large shallow depression occupied by the joint deltas of the Okavango and Chobe Rivers. The extent of the flooded area varies both seasonally and annually in response to seasonal and long-term drought cycles. When in full flood, its waters tend to overflow into the Makarikari depression, an extensive system of salt-pans covering about 3,000 sq. miles and constituting the northern edge of the Ngwato Reserve. (b) The Crocodile-Marico system has several subdivisions. The Motloutse basin, in the north, is topographically a large shallow saucer depressed towards the east; the drainage is fairly well defined, though many of the streams are seasonally intermittent. This area is inhabited mainly by Ngwato and Khurutshe. Farther south, an eastward extension of the Kalahari sands is associated with an absence of surface water and a consequent gap in human settlement. The ground then rises to the Limpopo-Molopo watershed, where surface drainage is best developed; this is reflected in a relatively high concentration of population both in the Transvaal and across the border in the Protectorate (Hurutshe, Kgatla, Kwena, etc.). (c) In the Molopo-Kuruman area the Rolong, Tlhaping, and Tlharo occupy the higher and moister south-eastern edge of the true Kalahari depression.

Except for the Okavango basin, Bechuanaland may be described as rolling country of insignificant relief, at an average altitude of some 3,000 ft. The eastern peripheries have rocky hills and ridges with well-defined river valleys offering opportunities for local water conservation and irrigation. Rock outcrops, associated with some of the oldest African formations, occur in the Transvaal and British Bechuanaland. Elsewhere the soils are derived from superficial deposits, and, although porous, are sometimes very fertile. In the Tawana Reserve (Ngamiland) the land accessible to flood waters is covered with a thick black flood alluvium, whose usefulness, however, is greatly diminished by its poor drainage.

Climatically two regions may be distinguished. The western region (British Bechuanaland and the Protectorate) is bounded by the rainfall isohyets of 10 ins. and 20 ins. The rains fall in the summer (October–April), usually in sharp, heavy showers that are extremely localized, and much of their benefit is also lost owing to the high rate of evaporation and the porous nature of the soil. The rainy season is preceded, during August and September, by a period of dry air, cloudless skies, and dust storms, which combine to create an oppressive weather. During the summer, days and nights are hot and unpleasant, but the winter is an exhilarating season with clear skies, warm sunny days, and cold nights. Winter frosts are experienced, and as they are extremely unreliable in their incidence they sometimes have an adverse effect upon the cultivation of crops. The eastern region (Limpopo valley and beyond) has a slightly higher rainfall (15–25 ins.), hotter and more humid summers, and less frequent incidence of winter frosts.

The vegetation, in the south, usually consists of grassland interspersed with thorn bush and patches of park. Northwards, the parkland increases until, by the time the northern limits are reached, it gives way to extensive dry thorn forest with occasional evergreens. Leguminous and bulbous plants are plentiful everywhere,

---

[1] Passarge, 1908, chaps. xv, xvi; Pim, 1933, pp. 1–6; Schapera, 1943 (a), pp. 1–7; Debenham, 1948, pp. 31–8; Pole-Evans, 1948; Sillery, 1952, pp. 195–9.

and, together with the fruits of certain trees, contribute to the food supply of the people. The Okavango basin is characterized by dense sudd and swamp vegetation. Eastwards, the grassland and parkland grade into " bushveld " with bush, tree, and luxuriant grass cover. The region as a whole is predominantly grassland, and good cattle country, although when over-grazed it quickly reverts to thorn-bush savannah. It is also, at best, marginal country for extensive cultivation; the unreliability of both frost and rainfall, combined with a general lack of surface water, make agricultural yields low and uncertain.

Before the coming of the Europeans, the whole region was also very well stocked with game of all kinds, and even today hunting is still possible in most parts of the Protectorate. Elephants, rhinoceros, and giraffe, formerly plentiful everywhere, have now disappeared except in the north-west, but hartebeest, wildebeest, gemsbok, eland, and kudu are still found in great numbers in all but the most settled areas. Smaller buck, especially duiker and steenbok, abound almost everywhere, together with hares and such game birds as korhaan, partridge, and guinea fowl. Jackals and various species of wildcat are hunted primarily for their skins, formerly made into karosses but nowadays usually sold to traders. The leopard, hyena, and wild dog present a constant threat to livestock, but lions, formerly numerous everywhere, are now found mainly in the western and northern districts, where they kill many cattle.

Land Tenure[2]

Formerly the Tswana were almost entirely self-supporting. They produced their own food, chiefly by raising crops and breeding livestock; they built their own homes, and made their own clothing and household goods, all from materials locally available. Cattle were often obtained by raids on weaker neighbours, but there was relatively little inter-tribal trade, except in times of crop failure. Contact with Europeans has introduced new commodities of many kinds, most of which are obtainable only by import. Their acceptance has greatly extended the range of wants, and created new standards of wealth and social status. Money has become the principal medium of exchange, and is needed universally for paying taxes and other dues. To satisfy all these wants, many Tswana have adopted new occupations, including especially wage-labour in European areas.

Despite the changes just mentioned, most Tswana still depend mainly upon the land for food and many raw materials. Each tribe formerly had its own territory, but today, as already mentioned, much of Bechuanaland is occupied by Europeans and other non-Africans. In the Protectorate, of the Tswana living under rural conditions, 98.6% are in Reserves demarcated for them by the Government, 0.9% on Crown Land, and 0.5% in European areas; in the Transvaal, 17% are in Reserves, 34% on land purchased tribally or privately, and 49% on European-owned farms; in British Bechuanaland 67% are in Reserves and 33% on European-owned farms; and in the O.F.S. 43% are in Reserves, 9% on privately owned Native farms, and 48% on European-owned farms.[3] Some 3,000 Tswana in the Transvaal, 300 in the Cape, and 2,500 in the O.F.S. live on privately owned land (acquired by purchase or otherwise) outside the main tribal areas;[4] a similar, but not identical, form of individual ownership obtains in the Protectorate in what are known as the " Barolong Farms " (Lobatsi District).[5]

In the Protectorate, almost every tribe officially owns the Reserve that it occupies. But it may not alienate land to non-Africans without the consent of the Government, which also has the power to regulate the use of the land, e.g., by

[2] Schapera, 1938, pp. 195–213 ; 1943 (a) ; 1943 (b) ; 1943 (c) ; 1945 (a) ; Language, 1941, pp. 292–307 ; Scroggie, 1946, pp. 43–50.
[3] The figures are for the districts listed above on pp. 11 f. and have been derived from the sources cited there.
[4] Cf. Rogers and Linington, 1949, pp. 102, 126–8. The figures cited are from the *Union Census Report*, 1946, Vol. I, Table 10.
[5] Schapera, 1943 (b).

prohibiting the cultivation of noxious plants, protecting roads and water supplies, preserving game and timber, and controlling the movements of livestock. Europeans and other non-Africans require the chief's permission to trade and carry on other business in his Reserve; they generally pay rent for their land, and have limited grazing facilities. The Government and missionary societies hold sites for their special purposes free of charge, but the land itself remains tribal property. In the Union all tribal land, whether owned by the Government (e.g., Reserves) or purchased by the people themselves, is controlled by the South African Native Trust, a Government agency established in 1936; its powers are similar to, but more extensive than, those of the Government in the Protectorate.[6]

Except for Government supervision, the use of tribal land is still controlled by the chief, acting through the headmen of villages and wards. They should see that every married man receives free grants of residential and cultivating land, which he, in turn, portions out among his dependants. So long as he maintains his home, and uses or proposes to use his arable land, he has exclusive rights over them, and on his death they normally pass to his heirs. He is free to give away or lend out some of his land, but he cannot claim payment in return; sale and rent of land are both unknown in tribal law (but have been adopted by some Tswana outside the main tribal areas in the Union). If he moves away permanently, or is banished from the tribe, he forfeits his rights and his land becomes available for re-allocation.

Pastoral land, in the larger Western tribes, is divided into administrative districts under overseers, whose permission is needed to keep cattle there. Elsewhere people may establish their cattle-posts wherever they wish outside the arable zones and villages. Natural waters and grazing are used in common, but any man sinking a well or building a dam has exclusive rights over the water it contains. Game and other natural resources are accessible to all. It was formerly taboo to cut certain species of tree early in the rainy season, and nowadays hunting, cutting wood, digging earth, etc., may be locally controlled in various ways; otherwise, anybody can hunt when and where he wishes, or take wood, grass, clay, earth, or edible wild plants, wherever he finds them.

PRODUCTION OF FOOD

Agriculture[7]

The principal crop in former times was Kafir corn (*sorghum vulgare*); maize, millet, sweet cane, earth-nuts, beans, and cucurbits were also cultivated, but seldom in great quantity. Nowadays more maize is planted, and many people also grow some tomatoes, potatoes, oranges, peaches, and similar innovations. There has been no other significant change in type of crop.

A few cereals, vegetables, and fruits are grown in very small quantities in backyards at home, but agriculture is carried on mainly at the fields (*masimo*). These are located outside the villages in compact blocks, often several square miles in extent, within which very many different families have adjacent holdings. The soils most favoured are sandy loams, suitable for a wide variety of crops; but their fertility, although initially high, is rapidly depleted by cropping, owing to inherent phosphorus deficiency and loss of organic matter. In the old days fields were seldom larger than two or three acres, but since the adoption of the plough the usual size has become between five acres and 20. Measurements in the southern Protectorate of 970 fields, embracing in all an area of 10,104 acres (average 10.4), showed that 424 (44%) were less than five acres each, 234 (24%) between five and 10 acres, 116 (12%) between 10 and 15 acres, and 194 (20%) over 15 acres.[8] In the larger

[6] Schapera, 1943 (a), pp. 35–40 ; Rogers and Linington, 1949, pp. 144–61.
[7] Passarge, 1905, pp. 686–9 ; Lebzelter, 1933, pp. 55 f. ; Pim, 1933, pp. 136–41 ; Schapera, 1943 (a), chap. vii ; Scroggie, 1946, chaps. v–viii.
[8] Schapera, 1945 (a), p. 27. The fields were located in the following Reserves : Kgatla (278), Tlôkwa (163), Ngwaketse (193), and Barolong Farms (336). The figures for the Barolong Farms were obtained by Mr. F. O. A. Wande, Agricultural Officer.

tribes most families have two or three fields each, often situated in different localities to overcome the hazards of an uncertain rainfall; in the smaller and more congested Reserves (e.g., Malete and Tlôkwa) few families have more than one, and up to 25% may have none at all.[9]

The cultivating season starts with the first rains, usually in November. Most families then move out to their fields, where they remain until the harvest, if any, is reaped in June or July. The men break the soil, using ox-drawn ploughs, and broadcast the seed by hand; the women and girls then do the weeding with hoes; and, when the corn starts seeding, they and the children are occupied daily in driving away the birds. The ears, when ripe, are harvested by hand or with knives. Threshing is done on a specially prepared floor of hard earth; oxen are occasionally used, but normally the women beat out the corn with heavy wooden flails. It is then winnowed, ashed to keep away weevils, put into bags, and transported on wagons or sledges to the family's home in the village. There it is stored in the granaries; in the east, these consist of large earthen bins kept in the open courtyard and mounted on stones for protection against white ants; in the west, similar bins or very large baskets are built on a raised platform inside a store-hut.

Formerly the cultivating season was everywhere inaugurated by **rain-making ceremonies**, which in times of drought were followed by more elaborate special usages.[10] These ceremonies were all organized by the chief. He also regulated the cycle of activity: people might not plant, weed, reap, or thresh for themselves until the work had been done on the " tribute fields " (*masotla*) cultivated for him by certain sections of the tribe. The normal rain-making ceremonies have now been abandoned, but in many tribes there is an annual Church " day of prayer for rain," originally introduced by the missionaries. Some also continue to employ professional rainmakers in times of drought. It is still a widespread rule, in the Protectorate, that ploughing may not start until the chief has given permission; seasonal taboos of various kinds are also observed to protect the crops; and some people employ magic on their own fields to keep away pests, counteract sorcery, and ensure an abundant harvest.

Apart from such practices, agriculture in general suffers from the use of poor seed, inefficient planting and weeding, and neglect of manure. The Governments have fairly recently placed trained demonstrators in many Reserves, introduced new varieties of seed, encouraged winter ploughing and rotation of crops, organized annual shows, and started vegetable gardens at the schools, where instruction is also given in the science of agriculture. Irrigation schemes, usually on a small scale, have also been started in the Ngwaketse and Malete Reserves (Protectorate) and in Taungs and Kuruman districts (Cape Province), and others are being planned. These attempts at improvement have on the whole had relatively little success; most people still produce less than one bag (200 lb.) of Kafir corn per acre, whereas demonstration plots have yielded more than 10 times as much. The uncertain rainfall remains a major handicap. Because of droughts, most fields are left uncultivated at least once in three or four years, and crop failures are frequent; in both British Bechuanaland and the Protectorate grain must often be imported to make good the deficiency.

Animal Husbandry[11]

Agriculture provides the Tswana with their staple food, but they themselves show more interest in animal husbandry. In pre-European times they kept cattle, goats, sheep, fowls, and dogs; they have since acquired horses and donkeys, and

---

[9] Schapera, 1943 (a), p. 135 ; Scroggie, 1946, p. 47.
[10] See below, p. 60.
[11] Language, 1941, pp. 328–39 ; Lebzelter, 1933, pp. 58–60 ; Passarge, 1905, pp. 690–3 ; Pim, 1933, pp. 117–36 ; Schapera, 1943 (a), chaps. xi, xii ; Schultze, 1907, pp. 629–35 ; Scroggie, 1946, chaps. ix–xv.

the Eastern tribes also have pigs. The distribution of cattle and small stock in tribal areas in the Union, at 31st August, 1948, was as follows:[12]

District	Cattle	Sheep	Goats	Animal Units	Units per Person
Kuruman	10,121	28,418	45,790	22,489	1.7
Mafeking	54,589	52,237	63,342	73,852	2.4
Taungs	22,547	47,150	66,594	41,504	2.1
Vryburg	23,981	25,062	41,078	35,005	1.9
Lichtenburg	19,501	29,650	10,057	26,102	1.8
Marico	35,766	5,611	10,611	38,470	1.7
Rustenburg	86,719	11,305	24,506	92,687	1.6
Thaba Nchu	12,043	4,977	—	12,782	0.9

Comparable district figures have not been published for the Protectorate, but at the time of the 1946 census the livestock holdings of the total Native population included some 628,800 head of cattle, 297,400 goats, 144,800 sheep, 2,700 horses, 21,400 donkeys, 4,600 pigs, and 127,800 fowls;[13] this gives a distribution of 2.4 animal units (cattle and small stock) per person.

Individual holdings are very difficult to ascertain, owing especially to the widespread custom of placing out cattle and small stock on loan.[14] A survey made of six tribal areas in the Protectorate in 1943, and embracing 4,047 families, showed that 298 (7.4%) had no cattle at all, and 749 others (18.5%) less than 10 head each, which is said to be the minimum size of herd that will yield an annual disposable surplus;[15] on the other hand, chiefs and a few other men have as many as 500–5,000 cattle each, or even more.[16] A survey subsequently made of 517 stock-owners in the Barolong Farms showed the following distribution of animal units per owner: 1–10, 18%; 11–20, 19%; 21–50, 40%; 51–100, 18%; over 100, 5%; average per owner, 38. But 57 of the men (11%) had no cattle at all, but only small stock, and, on the other hand, 12 (2%) had more than 100 head of cattle each, and another 49 (9%) between 51 and 100 head.[17]

Horses and donkeys have replaced oxen for riding, and dogs are used mainly in hunting. The meat of all the other animals is eaten when available, and the milk of cows and goats is drunk, both fresh and sour. Cattle, and sometimes sheep, are given as *bogadi* (bridewealth);[18] and oxen, formerly used as pack-animals, are nowadays harnessed to ploughs, wagons, and sledges. Cattle and small stock were also, and still are, standard mediums of exchange, and their sale for export is the principal local source of cash income. The possession of cattle is itself a source of status; a man's wealth is estimated mainly by the size of his herds, and a large owner is generally respected and influential in tribal affairs. By placing out his cattle on loan, he can also command the labour and allegiance of the holders for various purposes.[19]

Except when specially required elsewhere, cattle, small stock, and equines are kept at grazing-posts (*meraka*) in the open veld. Bechuanaland, on the whole, is

[12] *Report on Agricultural and Pastoral Production*, 1947–8 (U.G.30, 1950), Table 2. The number of animal units, and their *per caput* distribution, have been calculated from the published figures. An "animal unit" (according to the formula used by the Native Affairs Department, S. Rhodesia) = 1 head of cattle, 6 goats, or 6 sheep.
[13] *Census Report*, 1946, Table VII. (The number of goats is erroneously printed there as 29,409 instead of 297,409).
[14] See below, p. 28.
[15] Schapera, 1947 (a), p. 133.
[16] Schapera, 1933 (c), p. 650.
[17] Wande, 1949, Tables 11–13.
[18] See below, p. 41.
[19] See below, p. 28.

excellent ranching country, with many species of nutritious grass. In the old days, owing to scarcity of surface water, many good pasture areas could not be used during the dry season. Wells, bore-holes, and dams have now made some such areas permanently available, and thus increased the carrying capacity of the land; but cattle must still often be driven many miles from grazing to water, and sometimes they drink only once in two or three days. The larger Reserves, especially in the Protectorate, are not overstocked, but in some of the smaller ones (e.g., Malete and Tlôkwa) the problem is so acute that the people have had to buy more land or hire grazing from their neighbours.[20]

In the smaller Reserves, the cattle-posts are often near home; elsewhere, they are sometimes up to 100 miles away or even more. They usually consist of a hut or two for the herdsmen, and separate kraals for the mature cattle, calves, and small stock. The vast majority contain less than 100 animals each, and many less than 50. Each cattle-keeping family normally has its own post, or shares one with friends; wealthy men, however, may have several, often widely separated. Wealthy men also employ special herdsmen, but most cattle-posts are looked after by the adolescent sons of the owner, who visits them periodically to supervise.

The posts are not permanent fixtures. In the dry season they are generally concentrated near boreholes, dams, wells, or standing pools in the rivers, but when the rains come they are moved away, the cattle then drinking from pans and other open waters. The animals range freely over the surrounding pastures. During the dry season they are herded daily, for they need to be watered, and at night they are usually shut in the kraals to prevent them from straying; but in the rainy season they are left unattended. They are not stall fed, nor is fodder grown specially for them; after the harvest, however, they are usually taken to graze on the stubble in the fields.[21]

New water supplies (e.g., boreholes) are only one of the developments in animal husbandry due to contact with Europeans. Others include more efficient methods of treating disease, crossing local breeds of livestock with superior imported strains, and, in some localities, the commencement of dairy industries. In the Protectorate, " livestock improvement centres " have been established in many Reserves, and an elaborate system of control by the Veterinary Department includes the periodical inspection of every cattle-post.[22] Many people have taken profitable advantage of these developments; many others, however, still prefer quantity to quality in accumulating livestock, and also resort to magic instead of science for treating disease and promoting fertility.[23]

Hunting[24]

Although game has disappeared from the most settled parts of the country, it is still abundant in the Protectorate. Nowadays big game may not be killed without a permit, and several tribes have other local restrictions; but, almost everywhere, hunting remains an important pursuit. Men travelling in the veld usually carry guns or other weapons, both for protection and in order to kill for the pot. Boys at the cattle-posts pursue small buck with dogs and clubs, or set traps and snares for them and game birds. Professional hunters go into the veld for weeks at a time in the winter; they seek mainly for fur-bearing animals, but sometimes also hunt big game, whose meat they dry and sell in the villages. The chief occasionally organizes a large collective hunt, e.g., to destroy beasts of prey, and similar hunts are usually also held whenever an age-regiment is sent to round up stray cattle, or a *letsholô* meeting takes place out in the veld.[25]

[20] Cf. Schapera, 1943 (c) ; 1945 (a).
[21] Schapera, 1943 (a), pp. 217–22 ; Language, 1941, pp. 339–42 ; Scroggie, 1946, pp. 120 -7.
[22] Schapera, 1943 (a), pp. 212, 231.
[23] Schapera, 1934 (b) ; Scroggie, 1946, pp. 128–31.
[24] Passarge, 1905, pp. 694–7 ; Language, 1941, pp. 350–5 ; Schapera, 1943 (a), pp. 255–8 ; Scroggie, 1946, chap. xvi ; and, for legislative control of hunting, Schapera, 1943 (e), p. 45.
[25] See below, pp. 29, 53.

Food and Narcotics[26]

Formerly the staple food was Kafir-corn porridge, prepared in several different ways. The other cultivated plants were also eaten, and so were certain wild fruits, which if the crops failed often served to keep off starvation. Domestic animals provided milk, but they were seldom slaughtered, save by wealthy owners or on ceremonial occasions; meat was more usually obtained by hunting. In general, everybody ate the same kinds of food, except that wealthy people were better supplied, but minor distinctions were created by sex, age, and totemic, taboos. Nowadays exotic fruits and vegetables are also grown and eaten, and many people have acquired a taste for such innovations as tea, sugar, bread, preserves, and canned meats. In the main, however, food habits remain substantially the same. The principal meal is taken in the evening; there is no other regular time for eating, although food is usually also prepared in the morning.

Despite the apparent variety of the diet, malnutrition is common.[27] Except when living at their fields or cattle-posts, many people lack green vegetables or milk, their food consisting almost entirely of Kafir-corn porridge. Meat has also become scarcer owing to the diminution of game; and although it can now be bought from butchers in most of the big villages, it is so rarely eaten by some that they consider it a luxury.

Kafir corn is also converted into a mild and much-relished beer, especially on festive occasions. But many Mission churches forbid their members to drink it, and in some tribes the chiefs have either prohibited it altogether (e.g., Ngwato) or else control its sale and consumption (e.g., Ngwaketse). European liquors, e.g., brandy, were popular in the early days of contact, and some Western tribes later became addicted to *kgadi,* a fermented honey beer often blended with syrup; both are now officially prohibited everywhere.[28] Tobacco, both imported and locally grown, is used very extensively, and cigarettes have recently become popular. " Dagga " (Indian hemp) was formerly also smoked, but its cultivation and sale are now prohibited.

ARTS AND CRAFTS

Clothing and Ornaments[29]

Formerly, clothing was made chiefly from the skins of wild and domestic animals. Babies went naked; small boys wore a flap barely covering the genitals, youths and men a loin-skin passing between the legs and tied round the waist; small girls wore a tiny apron of fringes made of bast or skin, older girls and women a large apron in front and a skirt at the back. A kaross and sandals were often added, and men and boys sometimes wore skin caps or grass hats. Everybody, including babies, also had ornaments of some kind, such as necklaces, bangles, and anklets, made of beads, copper wire, or woven grass. Distinctive and more elaborate costumes were worn on occasion by initiands and warriors, but, on the whole, everybody of the same sex and age usually dressed alike, although karosses and ornaments varied greatly in quality and value.

The old costume is nowadays rarely seen, except among the younger children. " Respectable " dress is obligatory for Church members, but almost everybody else also wears clothes of European type and material. Women usually have at least a full blouse and skirt, with one or more petticoats, and, in cold weather, a blanket or shawl, and the standard costume for men is shirt and long trousers. Many people dress more elaborately, e.g., in full European costume, and some have several changes of clothing, for use according to the occasion. Ornaments are generally of the same type as before, but imported varieties are nowadays preferred.

---

[26] Squires, 1949, pp. 23–35 ; Passarge, 1905, pp. 698–700 ; Lebzelter, 1933, pp. 52–62 ; Schapera, 1940 (a), pp. 156, 159 f., 226 f. ; Scroggie, 1946, chaps. xxiv–xxxi.
[27] Squires, 1943, 1949.
[28] On tribal legislation about liquor, cf. Schapera, 1943 (e), pp. 36–8.
[29] Passarge, 1905, pp. 668–70 ; Lestrade, 1929, p. 20 ; Fritsch, 1872, pp. 168–72 ; Schultze, 1907, pp. 637–44 ; Hirschberg, 1936, pp. 27 f. ; Schapera, 1940 (a), pp. 47 f., 74, 123.

Dwellings and Household Goods[30]

The traditional type of hut consists essentially of a circular wall surmounted by a conical roof of thatched poles with protruding eaves; in the west, but not in the east, the roof is supported by a central post. Except among the Tawana, where it is made of reeds lashed closely together, the wall is of earth, sometimes plastered over a framework of canes; it has no windows, and the entrance is a gap which can be covered, if necessary, with a wickerwork frame. The floor is of beaten earth smoothed over with a mixture of cowdung and mud, and a shallow depression serves as a hearth. This type of hut is still often seen. Many modern huts, however, have good wooden doors, glass windows, and the European style of thatching. Progressive people sometimes build rectangular multi-roomed houses of European type; these are usually made of earth and thatch, but occasionally of brick and galvanized iron roofing.

Household utensils and implements formerly in common use included baskets, mats, beer-strainers, and winnowing-trays, all made of grass or reeds; clay pots for cooking, or for holding water and beer; iron-bladed hoes, axes, spears, and knives; skin bags, mats, and milk-sacks; wooden stools, food-bowls, clubs, milk-pails, stamping-blocks, porridge-stirrers, firesticks, and spoons; and calabash cups, bottles, and scoops. All these are still often seen, but many are being replaced more and more by such imported substitutes as metal ploughs, hoes, axes, spades, cooking-pots, buckets, basins, and cutlery, hessian grainbags, fibre suitcases, and various kinds of table-ware. Other imported goods in fairly great demand, some of which had no previous equivalent, include bedsteads, tables, chairs, lanterns, and sewing-machines, and such groceries as matches, candles, paraffin, salt, soap, and patent medicines.[31] New means of transport are represented chiefly by ox-drawn wagons and sledges; the latter are made locally, from the forked branch of a tree, and the former are imported (in 1946 the inhabitants of the Tswana districts in the Protectorate (excluding Ngamiland, for which data were not available) owned some 2,620). Nowadays the bicycle is also in common use, and some men even have motorcars and lorries.

TRIBUTE AND TAXATION

Apart from procuring for its own maintenance most of the goods mentioned above, every household owes certain forms of gift and tribute to others.[32] People always were, and still are, expected to give food to visitors; a selfish or niggardly person is generally despised. On occasion a man should also make special gifts to his close relatives, e.g., send them meat or beer whenever he slaughters or brews, contribute towards the *bogadi* for a wife, buy clothes or ornaments for a child being initiated or confirmed, and help to pay any tax, fine, or compensation due by someone with insufficient means.

The chief was formerly also entitled to various kinds of tribute from his subjects.[33] He claimed from hunters the brisket of every big-game animal they killed, one tusk of every elephant, and the skin of every lion or leopard. The head of every cattle-owning household gave him a beast at his installation, he received an ox from the father of every child being initiated and from the estate of every newly-deceased person of note, and, in years of good harvest, every woman sent him a basketful of corn. In some tribes he later claimed £1 or £2 in cash from every man returning home after working abroad, and he sometimes imposed an annual or *ad hoc* levy in cash upon all his men in order to finance education, water development, etc., or the payment of his own debts.

[30] Fritsch, 1872, pp. 172–88; Passarge, 1905, pp. 671–84; Schultze, 1907, pp. 623–5; Hirschberg, 1936, pp. 28–30 (technology); Lebzelter, 1933, pp. 50–3; Schapera, 1940 (a), pp. 95 f.; 1943 (a), p. 84.
[31] Lists of the principal commodities locally bought from traders or directly imported are given in Schapera, 1947 (a), pp. 228–32, 235–40.
[32] Schapera, 1938, pp. 239–41 and *passim*; 1943 (a), pp. 259 f.; Scroggie, 1946, chap. xviii.
[33] Schapera, 1938, pp. 63–6; 1943 (a), pp. 43, 155–7, 196–8, 258–62; 1943 (e), pp. 28 f.

Almost all the tribute formerly paid to the chief has now been superseded by annual taxation. In the Protectorate, every man aged 18 or over must pay from 5s. to £10 (according to his wealth in livestock or income from wages) to the local Tribal Treasury, and a further 28s. to the Government (which refunds 35% to the Tribal Treasury).[34] In the Union, there is a " general tax " of £1 per man, and a " local tax " of 10s. per hut, with a maximum of £2; the proceeds of the local tax, and one-fifth of the general tax, go to the local Native Council or, where this does not exist, to the South African Native Trust Fund.[35] In both territories, moreover, the tribal authorities with Government approval may impose *ad hoc* levies for some special purpose, e.g., building a school or buying land.

Many people have other forms of recurrent cash expenditure. In the Protectorate, for instance, cattle-owners pay special fees for watering their stock at tribal boreholes; money is often needed for Church dues, school fees and books, medical treatment, postage stamps, etc.; and it is the only form in which payment can be made for goods personally imported from abroad, or for transport by rail or bus.[36] Fines imposed in Government courts have always been payable in cash, which tribal courts too now use as an alternative to livestock.

ORGANIZATION OF LABOUR

Household Production[37]

Each household formerly produced the great bulk of its food, built its own huts and granaries, and did its own housework. Men, women, and children all contributed. Women tilled the fields, built and repaired the walls of huts, granaries, and courtyards, thatched roofs with grass which they fetched themselves, prepared food and made beer, looked after the fowls, fetched water, wood, and earth, collected wild edible plants, and did all the other housework; men herded cattle, hunted, did all the timberwork in building, cleared new fields, and occasionally assisted in planting, weeding, and reaping. Children helped according to age; they started in early youth with the simpler tasks, and by the time they were initiated did the same work as adults. Most of the livestock, especially, were herded by boys. This division of labour still persists in the main. But, now that oxen are used as draught animals, men do the ploughing, and sometimes also fetch wood, earth, etc., in wagons or sledges. Women are no longer debarred from handling cattle, and accordingly often assist in ploughing and driving; they still hardly ever herd or milk. In the east, they have also acquired the new task of looking after pigs.

Specialized Crafts[38]

Formerly each household also made much of its own clothing, ornaments, utensils, and implements. Men did all the work in skin, wood, metal, and bone, women made pots, and both sexes did basketwork, each making different kinds of objects. But some of these occupations were specialized. Metalwork and pottery were confined largely to certain families, within which the craft was handed down from parent to child; and although most men could work in skin, wood, etc., some were noted for the superiority of their products, or else made certain objects which others could not (e.g., food-bowls and karosses). Nowadays, owing to the spreading use of imported goods, fewer people than before make pots, baskets, wooden utensils, etc., and the art of working iron and copper is virtually extinct. Contact with Europeans, however, has created new skills. Building and thatching a hut in the European manner is an art practised by certain men only, and others specialize in making tables, chairs, etc.; some women, too, are professional dressmakers, although

[34] *B.P. Annual Report*, 1950, p. 9; Schapera, 1947 (a), pp. 123-5.
[35] *Official Year Book*, No. 25 (1949), p. 501.
[36] Schapera, 1947 (a), pp. 125-7.
[37] Brown, 1926, pp. 49-51; Schapera, 1940 (a), pp. 127-33; Scroggie, 1946, chaps. vi, xxiv.
[38] Lestrade, 1929, p. 21; Schapera, 1933 (c), p. 65; 1938, pp. 217, 241-2; 1940 (a), pp. 121, 135, 138, 338; Scroggie, 1946, chaps. vi, xii.

many others can make garments from purchased materials for the members of their household. Every specialist, incidentally, carries on the same farming activities as other people; the craft he also practises is merely a part-time occupation.

Labour Assistance[39]

Tasks that a household is too small to carry out alone, or wishes to complete reasonably soon—e.g., clearing a new field, threshing corn, building the wall of a hut, or thatching a roof—are generally done with the aid of relatives, or by organizing a work-party (*letsema*) and paying with meat, beer, etc. the neighbours coming to help. Men with no suitable young relatives employ outsiders to look after their cattle-posts. Wealthy men place out cattle and small stock on loan; this both simplifies the task of herding, and insures against total loss from disease, etc. The holders of such " loan cattle " (*mafisa*) often render other services to the owner. In return, they use the oxen for ploughing and transport, take the milk of the cows, and usually receive a heifer for themselves. Men locally engaged in such occupations as teaching and trade sometimes hire others to do work for which they themselves cannot find time, e.g., clearing new land, digging wells, or cutting and transporting firewood, rafters, and thatching-grass. Certain tasks, too, are performed only by specialists, who must be hired by households needing their services; thus, magicians are widely employed to " doctor " homes, fields, and cattle-posts, and people wanting modern huts employ skilled builders and thatchers.

Retainers and Serfs

Chiefs and other men of rank have hereditary retainers.[40] These people usually consist of one or more family-groups descended from impoverished tribesmen or refugees, and belong to the same ward as their master;[41] the chief, however, is served by several different wards, whose original headmen were specially selected for their loyalty and devotion. Some of the retainers work continuously in their master's home or at his cattle-posts, and the group as a whole also cultivates his fields, builds his huts, and does whatever else he may require. Nowadays he sometimes pays his full-time servants in cash (e.g., 5s. to 10s. per month), but he generally follows the old system of feeding and clothing them, letting them use his cattle and ploughs, and when necessary helping them with tax or bridewealth. Their relatives share incidentally in these benefits. Individual members of the group are free to seek other work if they wish, but the existing arrangement seems to satisfy most of them.

In the Western tribes, chiefs and other prominent men formerly also had serfs (*malata*).[42] The master at first merely claimed tribute in kind, whatever his serfs produced or acquired being at his disposal; subsequently he gave them dogs and firearms with which to hunt for him; and, still later, he also employed them for menial work at his cattle-posts, fields, and home. Unlike his retainers, they were not free to move away and work for someone else, although he himself could give or lend them to other people; and if they ran away, as they sometimes did, they were usually pursued and brought back by force. Compulsory servitude of this kind has now been legally abolished, and in principle the former serfs have the same rights and privileges as other retainers.

Regimental Labour[43]

Large-scale activities are generally organized through the age-regiments.[44] Once formed, a regiment may be called upon at any time for public service. The men's

---

[39] Schapera, 1938, pp. 246–8, 253–6 ; 1940 (a), pp. 132 f. ; Scroggie, 1946, chaps. vi, xii.
[40] Schapera, 1938, pp. 66–8, 248–53 ; 1940 (b), pp. 77 f. ; Ashton, 1937, p. 77 ; Language, 1943 (a), pp. 27–31 ; Scroggie, 1946, pp. 19 f., 64 f.
[41] For the " family-group " and " ward," see below, pp. 40, 46–47.
[42] Tagart, 1933 ; London Missionary Society, 1935 ; Joyce, 1938 ; Passarge, 1905, pp. 710 f. ; Schapera, 1943 (a), pp. 27, 30, 260–2 ; 1943 (e), pp. 28 f.
[43] Pim, 1933, pp. 107–10 ; Language, 1941, pp. 451–3 ; Schapera, 1938, pp. 109–12 ; 1940 (b), p. 74 ; 1947 (a), p. 120.
[44] See below, pp. 38–9.

regiments formerly constituted the tribal army and fought its wars; and, in times of peace, they were often employed to round up stray cattle, destroy beasts of prey, go on hunting expeditions, clear new fields for the chief, build his huts and cattle-kraals, search for missing persons, etc. With the abolition of inter-tribal warfare, their military functions have largely disappeared, although in both World Wars regiments from various tribes in the Protectorate were sent abroad by their chiefs to serve in the British Army. Their other duties still remain. In addition, they have been used for such new tasks as making roads and dams, helping to build schools and churches, and, in some tribes (e.g., Kgatla) seeking work abroad to earn money for public purposes. The women's regiments are similarly employed, e.g., to keep the village clean, fetch water and earth for building operations, and smear the walls and floors of the chief's compound. Only the chief can summon a whole regiment to work, but headmen of villages and wards may use their own detachments for local purposes. In the Protectorate regimental labour was until very recently compulsory, failure to participate when summoned being a penal offence in tribal law. The occasions and conditions of its use have now been statutorily defined, and people may also be exempted by paying a commutation fee.

Trade and Exchange[45]

Formerly there was much internal trade, both in iron implements, pots, karosses, and other objects made by specialists, and in livestock, corn, meat, and tobacco. These were all bartered for one another, and in some instances their relative value was stabilized (thus, hoes, spears, and axes were worth a goat each, and a pot its content in grain). There were no professional traders, nor any markets where goods were offered for sale; whoever wished to sell or buy something inquired among his neighbours until he found a customer, and the transaction was then concluded directly between them. In times of crop failure grain was often sought in other tribes, but normally there was little inter-tribal trade; the visitors, in such cases, had to obtain the local chief's permission and pay him something (e.g., a heifer or some rolls of tobacco) " to open the gate " for them.

Livestock and food are still often obtained from fellow-tribesmen by barter, although in most large villages there are now also butchers, bakers, and eating-house keepers, who usually demand payment in cash. Some household goods, too, are still supplied directly by their makers; but many can be obtained only by purchase from the trading-stores, which carry imported stocks. Until recently the traders were all Europeans or Asiatics, but tribesmen have now begun to operate stores of their own.

The traders also purchase local produce for export. The commodities chiefly in demand consisted at first of ivory, ostrich feathers, and the skins of fur-bearing animals, and nowadays of livestock, hides and skins of domestic animals, dairy products, Kafir corn, beans, and maize. Cattle are by far the most important, although exports from the Protectorate have often been restricted by weight embargoes and other special conditions. A good trade also exists in hides and skins, but poor methods of preparation have tended to keep prices low. Sheep, goats, pigs, poultry, and eggs have a steady but comparatively small sale. The production and marketing of cream has become a fairly flourishing industry in British Bechuanaland and in the north-eastern Protectorate. The trade in crops fluctuates greatly. In some years they figure prominently as exports; in others they are imported in large quantities because of extreme local shortages due to drought. In the Protectorate, moreover, most chiefs long ago forbade the sale of Kafir corn without special permission, experience having shown that in bad seasons people were often forced to buy it back, at much higher prices than paid to them by the traders. Nowadays all

[45] Passarge, 1905, pp. 700–3 ; Schultze, 1907, pp. 646–9 ; Schapera, 1933 (c), pp. 646–51 ; 1938, pp. 241–4 ; 1940 (a), pp. 138–41 ; 1947 (a), pp. 130–6 ; Scroggie, 1946, chaps. xix–xxii.

corn for sale must be offered first to the Tribal Treasuries, and only when an ample reserve is available can the surplus be sold, under permit, to traders.[46]

The average annual value of imported goods purchased by Africans in the Protectorate, for the years 1938–42, was approximately £460,000. During the same period their annual income from exports was about £242,500, of which £214,000 came from the sale of cattle.[47] But, as already mentioned,[48] a survey conducted in 1943 showed that one family in every four either has no cattle at all, or else so few that it cannot afford to sell a beast every year to meet its tax and other cash obligations.

WAGE-LABOUR

Local Employment[49]

To supplement their income from produce, many Tswana nowadays work for wages. Owing to insufficient demand, relatively few are so employed inside the tribal areas; of these, some work for fellow-tribesmen or European residents as cattle-herds, domestic servants, shop hands, etc., and some engage in teaching and other forms of Government service, including the clerical and technical branches of tribal administration. In the Protectorate, in 1943, some 3,000 males and 380 females were working locally for the Government, tribal administrations, and European residents; the number employed by fellow-tribesmen was not ascertained. Their cash wages ranged on the average between 17s. 6d. and 40s. per month (according to occupation), and amounted in all to about £82,000; most of them also received food and accommodation. Comparable figures for the Union are not available.

Labour Migration[50]

The great majority of wage-earners are employed away from home in European industrial and farming areas. Labour migration is a phenomenon at least 80 years old, but since about 1930 it has become much more common. Figures for the Union are not available, although the fact that there are only 82 males for every 100 females in the tribal areas suggests that the proportion of migrants is fairly high.[51] In the Protectorate, at the time of the 1946 census, 7.9% of the *total* male population and 2% of the *total* female population in the Tswana districts were recorded as being away from home, although many had merely gone visiting.[52] Data collected in 1943 showed that about 28% of all *adult* males may be working abroad simultaneously; the average varies from 40% in the south-east (Kwena, Ngwaketse, Kgatla, Malete, Rolong, Tlôkwa) and 26% in the north-east (Ngwato, Khurutshe) to as little as 6% in the north-west (Tawana). About nine-tenths of the migrants belong to the able-bodied age groups (15–44) and, in general, more bachelors are away than married men. Women also have recently begun to go abroad, although in much smaller proportions; the average (in the east) is 5%, of whom about one-fifth are together with their husbands.[53]

Of labour migrants from the Protectorate, about 9% go to other parts of the Territory, 89% to the Union (74% to the Witwatersrand, 10% elsewhere in the Transvaal, and 5% to other provinces), and 2% to Southern Rhodesia and elsewhere.[54] Of those going to the Union, about 60% work on the Witwatersrand gold mines; most of the others are unskilled labourers in commerce or industry, and relatively few are domestic servants or farm hands. Most men make several trips abroad before finally settling down at home. Mine labourers, almost all of whom

[46] For these and other trade restrictions, cf. Schapera, 1943 (e), pp. 47–9.
[47] Schapera, 1947 (a), pp. 129, 132 f., 234.
[48] See above, p. 23.
[49] Schapera, 1947 (a), pp. 136–9.
[50] Pim, 1933, pp. 29–32; Schapera, 1947 (a).
[51] See above, p. 13.
[52] *Census Report*, 1946, Table II (h).
[53] Schapera, 1947 (a), pp. 39, 40, 65, 68.
[54] Ibid. p. 46.

are recruited, usually stay away for about a year at a time; those in other occupations may be away uninterruptedly for two or three years, and sometimes much longer. About 6% have been away so long that they are now considered lost to their tribes.[55] The Native Labour Proclamation of 1941 provides for the repatriation of all recruited labourers, but this does not affect the large proportion (about 40%) who go abroad independently. The Proclamation also prohibits direct or indirect recruiting by Government officials or chiefs and headmen, specifies the conditions under which labour agents may be licensed and carry on recruiting activities, empowers the Resident Commissioner to limit the number of men recruited from any district, and provides for written contracts of employment (which must be attested before a Government officer), medical examination of recruits, and their transport to the place of employment.[56]

The export of labour has become an outstanding feature of the modern economy. Without the income that it produces, the Tswana could not possibly maintain their present standards of living. Thus, it has been estimated that about £54,000 is earned annually by people working in the European areas of the Protectorate, and that from the wages of those going abroad about £333,000 comes back either in cash or in the form of goods; together, these constitute 54% of the total income (the balance being derived from the sale of produce and from employment inside the Reserves).[57] On the other hand, farming and other local activities have been adversely affected by the drain upon domestic labour resources, the stability of the family is being weakened by the prolonged absence of husbands and the associated infidelity of wives, and the men on their return show less respect for traditional authorities and obligations.[58]

Tribal Finances

Formerly, as already noted,[59] the chief received tribute from his subjects in cattle, corn, wild animal skins, ivory, and ostrich feathers; he also kept most of the cattle looted in war, all unclaimed stray cattle, and most of the fines imposed in his court. In return, he gave meat and beer to people visiting him, assisting at his court, or summoned to work for him, and rewarded with gifts of cattle the services of his warriors and councillors; he placed many of his cattle as *mafisa* with poor men, who lived upon the milk and ploughed with the oxen; and in times of famine he provided the people with corn from his granaries, or purchased supplies to distribute as food and seed.[60]

Since the coming of the Europeans, cattle-raiding has ceased, and hunting tribute has diminished greatly in value. But for many years the chief enjoyed new sources of income. Thus, in the Protectorate, he received an annual commission of 10% on all Government tax paid by his people, from time to time imposed levies in cash for public purposes, and had at his disposal the stand rents paid by traders and (in some tribes) the subsidies paid for prospecting and wood-cutting concessions in his Reserve. The money accruing from these and other sources he used as he pleased. From about 1920 onwards, however, the Administration began to establish local Tribal Funds, each administered by a Board of Trustees with the District Commissioner as chairman and several other members chosen by the chief. Tax commission, levies, stand rents, etc., were all paid into the Fund; from it the chief received a fixed annual salary, and the balance was spent on economic and social developments, including education.[61]

[55] Schapera, 1947 (a), pp. 53–64.
[56] Ibid. pp. 101–12.
[57] Ibid. 157–62.
[58] Ibid. pp. 162–93.
[59] See above, p. 26.
[60] Schapera, 1938, pp. 68–9 ; 1940 (b), pp. 77–8.
[61] Pim, 1933, pp. 102–7 ; Schapera, 1938, pp. 65 f. ; 1940 (b), p. 78 ; 1947 (a), pp. 124 f., 152, 154 f.

In 1938 a Tribal Treasury was established in each Reserve to control all public revenue and expenditure. The Treasury is supervised by the Government, which audits its accounts, but it is operated by a small Finance Committee elected by the tribe or appointed by the chief. The Committee prepares the draft estimates, which are considered by the tribe and then submitted to the High Commissioner for approval; the routine work is done by a paid treasurer and other African assistants, including clerks and revenue collectors. The principal sources of revenue are graded tax, Native tax (35% of the total collection), fines imposed by the tribal courts, stand rents, interest on fixed deposits, commission on stock sales, water rates, sales of school books, and school fees, and, in some tribes, sales of grain and of cattle bred in the livestock improvement centre, sales of vaccine, dipping fees, hunting permits, mineral concessions, etc. Estimates for 1952–3 vary from £2,195 (Tlôkwa) to £42,000 (Ngwato). Expenditure includes salaries and wages of Native authorities and other tribal officials and employees (e.g., teachers, mechanics, police, social and agricultural workers, and unskilled labourers), capital and maintenance costs of school buildings, tribal offices, livestock improvement centres, boreholes, village water supplies, mechanical transport, etc., and the cost of building and maintaining roads.[62]

The following examples of revenue and expenditure for individual tribes come from the approved estimates for 1952–3:[63]

	Tlôkwa	Kgatla	Kwena	Tawana
	£	£	£	£
Revenue:				
Native Tax	244	2,000	4,500	3,150
Graded Tax	234	2,000	4,200	3,400
Native Courts: Fines	35	300	125	400
Tribal Levy (Arrears)	2	30	50	3
Sale of school books	30	125	200	90
Interest on fixed deposit	34	105	750	180
Stand rents	3	298	493	660
Stock sales commission	80	600	500	—
School fees	70	235	—	—
Water rates	—	150	450	—
Hunting permits	—	40	50	100
Sale of grain	15	—	2,750	—
Sale of livestock	—	726	—	—
Sale of arts and crafts	—	40	50	—
Sale of vaccines	10	—	—	60
Dipping fees	10	50	—	—
Concession premium	—	—	225	—
Ammunition permit fees	—	—	—	30
Government grant-in-aid	425	—	—	—
Education levy	1,000	—	—	—
Miscellaneous	3	20	75	50
	£2,195	£6,719	£14,418	£8,123

[62] *B.P. Annual Report*, 1938, p. 51; Ashton, 1949, pp. 714 f.
[63] Government Registry (Mafeking), File 4041.

	Tlôkwa	Kgatla	Kwena	Tawana
Expenditure :	£	£	£	£
Administration	270	1,600	3,049	3,407
Agriculture and Veterinary	20	780	2,730	100
Education	792	3,720	5,454	2,682
Judicial	—	—	15	10
Medical	—	62	76	49
Miscellaneous	52	163	533	188
Police	—	88	281	339
Works staff	36	72	624	251
Works recurrent	25	180	953	630
Works extraordinary	1,000	390	250	350
	£2,195	£7,055	£13,965	£8,006
Estimated deficit	—	246	—	—
Estimated surplus	—	—	453	117

There is as yet nothing in the Union similar to the Tribal Treasury, but some of the Transvaal tribes have for many years had special Trust Funds for the purchase of land and for such developments as fencing, constructing and maintaining dipping tanks, sinking boreholes, erecting windmills, etc. The revenue of the Fund is derived from an annual rate payable by each taxpayer; the proceeds are deposited in a special banking account, which is administered by a Board consisting of the tribal authorities and a local representative of the Native Affairs Department.[64]

In addition, there are Local Councils usually embracing several different tribes. Such Councils, for which provision was made in the Native Affairs Act (No. 23 of 1920, as amended), were established at Mafeking (1927), Moiloa Reserve in Marico District (1929), Rustenburg (1940), Taungs (1940), and Kuruman (1948). They have statutory powers to raise revenue by means of annual rates, to spend the money on developing and maintaining social and economic services, to acquire and hold land and interests in land, and to make by-laws in regard to any matter with which they are concerned. Each consists of not more than nine Native members, who serve for two years; some are appointed by the Government, and the others are elected by the taxpayers concerned. The local Native Commissioner presides at the meetings, held every two months, and acts as adviser, and the accounts are audited by the Government; estimates of revenue for 1950 range from £1,400 (Kuruman) to £3,930 (Rustenburg), with corresponding expenditures of £658 and £5,010 respectively.[65]

---

[64] *Official Year Book* No. 25 (1949), pp. 485 f. ; Rogers and Linington, 1949 ; pp. 85 f.
[65] *Official Year Book* No. 25 (1949), pp. 481, 483 ; Rogers and Linington, 1949, pp. 63–70.

SOCIAL ORGANIZATION

The Tribe[1]

The Tswana are divided into more than 50 separate tribes. Each tribe (*morafe, setšhaba*) is politically independent of the rest, and, subject nowadays to the over-riding control of the European Governments, manages its own affairs under the leadership and authority of a chief (*kgosi*); it has its own name, derived usually from some former chief or the traditional founder of the present dynasty, but sometimes from the totem of the royal family, the site of an ancient capital, or some historical incident;[2] and it usually also has its own territory, all people domiciled there being regarded as its members.

The frequent emergence of new tribes has been one of the outstanding features in Tswana history.[3] Almost every tribe claims to have originated by seceding from another, its leader being a member of the local royal family; no instance is recorded of a new tribe having been founded by a commoner.[4] But some of the tribes that formerly existed (e.g., Kaa, Mmanaana-Kgatla, and Sikô) no longer survive as politically separate entities; they were broken up by civil wars or invasions, and their people now live as subject communities in other tribes.[5]

Tribes vary greatly in size. Detailed figures are not available for all, especially in the Union, but the following will give some idea of the range of variation in the Protectorate: Ngwato 101,000, Kwena 40,000, Tawana 39,000, Ngwaketse 39,000, Kgatla 20,000, Malete 9,500, Khurutshe 2,900, and Tlôkwa 2,300. The size of the tribal territory, and the density of population, of the first five have already been noted;[6] the territory of the Malete is 178 sq. miles in extent (density 53.4), of the Khurutshe 161 sq. miles (density 18.0), and of the Tlôkwa 67 sq. miles (density 34.3).

Membership of a tribe is determined firstly by descent: normally a man belongs to the same tribe as his father, and remains there for life. But every tribe has a population of mixed origins. Even the Tlôkwa, smallest of those named above, include several groups differing in stock from the nuclear community (the people whose ancestors founded the tribe); and, at the other extreme, more than five-sixths of the Ngwato tribe consists of what were originally alien peoples.[7]

The members of a tribe sometimes also differ in customs and language. This feature is least marked in the Eastern tribes, where, except for a few recent immigrants, almost all the people are now culturally and linguistically homogeneous. But in the southern tribes of the Protectorate, especially Ngwaketse and Kwena, there are many groups of Sarwa and Kgalagadi, who still retain their distinctive characteristics. In the North, the diversity is very much greater, and several different languages are in current daily use. Thus, among the Ngwato, there are relatively large numbers of Sarwa (Bushmen), Birwa and Pedi (Transvaal Sotho), Kalaka (Shona), Rotse (Central Bantu), Herero and Koba (Western Bantu), and Kgalagadi, all of whom speak their own languages and have many usages different from those of their rulers. Similar conditions obtain among the Tawana, Khurutshe, and Seleka-Rolong (of Francistown district).[8]

The alien groups in a tribe were sometimes annexed by force, or else submitted voluntarily, when their territory was invaded and occupied; others came as refugees

---

[1] Schapera, 1938, pp. 2–7 ; Breutz, 1941, pp. 9 f.
[2] Examples of the different types are given by van Warmelo, 1935, p. 96.
[3] See above, pp. 15 f.
[4] For a discussion of tribal origins, cf. Schapera, 1952 (b), pp. 8–19.
[5] Schapera, 1945 (b), pp. 116–8 ; 1952 (b), pp. 43, 54.
[6] See above, p. 11.
[7] For details, cf. Schapera, 1952 (b), pp. 31–124, especially pp. 116–20 (Tlôkwa) **and pp. 65–93** (Ngwato). Similar details are given for the Tlhaping in British Bechuanaland **by Language,** 1943 (a), p. 78.
[8] Cf. Schapera, 1952 (b), especially pp. 127 f.

from an invading enemy or from the oppression of a conqueror, or seceded from their own tribe because of internal dispute. It is still fairly common for malcontents or victims of injustice to change their allegiance, and for men disputing the authority of their chief to be banished. Once accepted by the chief to whom they then apply, they become members of his tribe and are allotted a place in his territory. The tribe, therefore, is not a closed group, membership of which is permanently fixed by birth; it is, rather, an association into which people may be born, absorbed by conquest, or enter of their own accord, and from which, again, they may depart voluntarily or be expelled.[9]

Totemism[10]

The Tswana are also divided into approximately 25 groups,[11] each having a certain species of animal as its *serêtô, seanô, seila,* or *sebôkô* (object of honour, veneration, avoidance, or praise).[12] These groups cut across the division into tribes. Some (e.g., those with the crocodile, elephant, ape, or lion as their totem) are represented in almost every tribe; on the other hand, every tribe also contains members of many different totem groups (e.g., 18 such groups are represented among the Tshidi-Rolong, 16 among the Malete, and 22 among the Ngwaketse). People with the same totem are assumed to have had a common origin, and all those with different totems from their tribal chief are accordingly held to be of foreign extraction. The assumption is not always correct, since instances are known of people having changed their totem, usually for magical or political reasons; some sections of Hurutshe, for example, have had three successive totems (eland, baboon, and hartebeest).[13]

Totem groups are patrilineal but non-exogamous. There are special myths telling how each group acquired its totem, and people should not kill or eat their totem, nor touch its skin; should they do so inadvertently, they must undergo a purification ceremony to avoid illness or other misfortune. It is also considered polite to address or greet a person by the name of his totem. Apart from these usages, and the belief in a common origin, there is no special bond among the members of a totem group. Even the usages are not always taken seriously nowadays, and the taboos are sometimes impossible to violate, the animals concerned being locally extinct.

Territorial Organization[14]

The Tswana live in large compact settlements. In the old days, judging from historical traditions and the writings of early observers, it was fairly common to find all the members of a tribe concentrated in one village round their chief. The subsequent multiplication of settlements came about partly through the conquest of territory in which other peoples were already living, and partly through the accession of large groups of immigrants whom it was considered politic to place some distance away. Even now the nuclear community of almost every tribe is found mainly in the chief's village, the others being inhabited largely if not exclusively by conquered or immigrant peoples.

In the Protectorate, all the Tlôkwa are concentrated in a single village, and the Malete have only three (the capital, with 8,250 inhabitants, and two others, with

[9] Schapera, 1938, pp. 118–23 ; 1952 (b), pp. 21–7 ; Language, 1943 (b), pp. 78–88.
[10] Stow, 1905, pp. 408–17 ; Willoughby, 1905, pp. 295–301 ; Frazer, 1910, Vol. II, pp. 369–77 ; Brown, 1926, pp. 27–42 ; Lestrade, 1929, p. 12 ; Schapera, 1938, pp. 6 f. ; 1946, pp. 18 f. ; Breutz, 1941, pp. 10–12.
[11] This figure excludes the totems of non-Tswana peoples (such as those found in the Western tribes).
[12] The ruling sections of the Rolong tribes have two totems, the kudu and iron ; the latter is one of the very few non-animal totems.
[13] Willoughby, 1905, pp. 300–1. Other examples are given by Schapera, 1952 (b), pp. 40, 41, 43, 54, 55, 69, etc.
[14] Ashton, 1937, pp. 72–6 ; Schapera, 1938, pp. 7–12 ; 1943 (a), pp. 23–6, 59–68 ; Language, 1943 (a), pp. 3–10 ; Mackenzie, 1871, pp. 365–9.

populations of 730 and 520 respectively); in the other tribes, the capital is also by far the largest settlement, with populations ranging from 12,000 (Kgatla) to 23,000 (Ngwaketse), and all the other villages inhabited chiefly by peoples of Tswana stock contain several hundreds or even thousands of inhabitants. Thus, among the Kgatla, there are altogether nine villages outside the capital, their populations being 250, 580, 590, 870, 940, 960, 980, 1,080, and 1,420 respectively.[15] The Transvaal tribes have a similar system. In British Bechuanaland, and on the " Barolong Farms " in the south-eastern corner of the Protectorate, the large villages described by early writers are now rarely seen, most of the people being dispersed all over the country in small household settlements often a mile or more apart. Small settlements are also very numerous among the northern and western tribes of the Protectorate, but here their inhabitants are all peoples of non-Tswana stock.

In appearance the Tswana village is typically a cluster of small circular hamlets separated from one another by narrow lanes or broad roads.[16] Each hamlet is inhabited by a single ward or sub-ward,[17] and their number varies with the size and organization of the population. Villages are normally situated on the banks of rivers or at other places where water is readily accessible, and preference seems to have been given to sites where hills afford protection against enemy raiders. In the old days the people often moved because of war, famine, etc., but most villages, especially the capitals, have now been inhabited continuously for many years, e.g., Kanye (Ngwaketse) since 1853, Mochudi (Kgatla) since 1871, Ramoutsa (Malete) since 1875, and Mošaweng (Tlôkwa) since 1887.

In the larger tribes of the Protectorate, all the villages are nowadays grouped for administrative purposes into districts, ranging in number from seven (Ngwaketse) to eleven (Ngwato). Each district, except that of which the capital itself is the centre, is administered by a resident governor appointed by the chief.[18] The system was started among the Ngwato by Chief Kgama III (1875–1923), as a means of controlling more efficiently the people in the remoter parts of his territory; it was gradually imitated elsewhere, but was not fully developed until after the Government reforms of 1934–43.[19] The Tshidi-Rolong and Tlhaping in British Bechuanaland have a similar system, but its history has not been recorded. It does not exist among the smaller tribes of the Protectorate, where outlying villages can be supervised directly from the capital; and is apparently also (and for the same reason) unknown in the Transvaal.[20]

SOCIAL STATUS

Rank and Social Classes[21]

The varied origins of the population are reflected in many tribes (but apparently not all) by the existence of three separate classes: " nobles " (*dikgosana*), agnatic descendants of any former local chief; " commoners " (*badintlha, batlhanka*), descendants of aliens incorporated long ago; and " immigrants " (*bafaladi, baagedi*), people of groups more recently admitted. The first two are generally regarded as " true " members of the tribe. Immigrants, especially if not of Tswana stock, are as it were still on probation, but, if they remain, ultimately become accepted as commoners. The class distinctions operate mainly in political life. The chiefship itself is confined to the senior family of nobles; the tribe's inner councils usually consist of select nobles and commoners; and although all three classes participate in

[15] Extracted from district returns for the 1946 census.
[16] For descriptions of Tswana villages, cf. Mackenzie, 1871, pp. 365 f.; Passarge, 1905, pp. 676–8; Schapera, 1943 (a), pp. 68 f.; Scroggie, 1946, pp. 2–6.
[17] See below, p. 46.
[18] See below, p. 54.
[19] See below, p. 51.
[20] Ashton, 1937, pp. 70–4; Schapera, 1938, pp. 9–11, 96–101; 1943 (a), pp. 30 f.; Language, 1943 (a), pp. 15–25.
[21] Passarge, 1905, pp. 710–2; Ashton, 1937, pp. 70, 75, 76–8; Schapera, 1938, pp. 30–3, 66–8, 72–4; 1940 (b), p. 59; Breutz, 1941, pp. 71–4; Scroggie, 1946, pp. 15 f., 22 f.

the general assemblies, immigrant speakers seldom command as much influence as others.

Within each class there are further distinctions. Among nobles, the more closely a man is related to the chief, the higher does he rank. Among commoners and immigrants, status is determined mainly by such factors as political office and family connections; the head of any group is senior to all his dependants, among whom his own relatives are of higher status than others. In some tribes (e.g., Kgatla, Malete, and Tlôkwa), the heads of " sections " and " wards "[22] are also ranked in relation to one another, but this does not seem to be universal. The order of precedence was formerly always observed in such tribal ceremonies as initiation and eating the first fruits,[23] and even today, whenever men assemble for work or formal discussion, the most senior one present is accepted as leader.

The system of ranking is not rigid: commoners of lowly position, who are conspicuously loyal to the chief and able in other ways, may receive promotion by being made headmen of new wards. In addition, men with outstanding personal qualifications of special kinds (e.g., bravery in war, skill in debate, knowledge of the law, or proficiency in magic) usually gain prestige and therefore influence in public affairs; wealthy men, similarly, may obtain personal followers by lending out cattle. In modern times such people as Church officers, teachers, and the chief's clerical assistants have also acquired special authority, regardless of their hereditary status, and educated people in general play a prominent part in tribal affairs everywhere.[24]

In the Western tribes there was formerly also a class of " serfs " (*malata*), consisting mainly of Sarwa and Kgalagadi, but also (in the north) of Pedi and Koba.[25] These people, found in the country when it was occupied by the Tswana, were parcelled out in local groups among the chiefs and other leading tribesmen. They and their descendants were permanently attached to the families of their masters, to whom they paid special tribute and whom they served in various menial capacities; such property as they acquired was at their master's disposal; if oppressed, as they often were, they had no access to the tribal courts; and they lacked many other civic rights, including participation in the political assemblies. Most of their obligations and social inequalities have now been abolished, and they are usually classed as commoners; but some, especially the Sarwa, are still considered inferior to other members of the tribe, who deem it degrading, for instance, to intermarry with them.

Sex and Age Differentiation

Social distinctions of various kinds also exist between men and women.[26] They sit apart at feasts and other social gatherings, and certain spots in the village, like the *kgotla* (council-place), are normally reserved for men. Sons are preferred as children, and a woman bearing daughters only is often despised. There is a well-defined division of labour between the sexes, certain tasks being traditionally allotted to each. In tribal law women are treated as perpetual minors, being subject for life to the authority of male guardians; they are also excluded from political assemblies, and although a few have recently acted as regents during the minority of a chief (e.g., among the Ngwaketse and Tawana), all political offices are normally confined to men. Formerly they were likewise debarred from officiating at sacrifices and other religious ceremonies now no longer observed; there were, however, and still are, some female magicians, although this is essentially a male occupation.

[22] See below, p. 46.
[23] Willoughby, 1928, pp. 227–9; Ellenberger, 1937, p. 62; Language, 1941, pp. 324–5; Breutz, 1941, pp. 98–102; Schapera, 1946, p. 24.
[24] Schapera, 1938, pp. 32–4.
[25] In addition to the references given above, p. 28, cf. Mackenzie, 1871, pp. 128–34; Language, 1941, pp. 84–92; Breutz, 1941, p. 73 f.
[26] Schapera, 1938, pp. 28 f. and *passim*; 1940 (a), pp. 100–7, 336, 340, and *passim*; Breutz, 1941, pp. 107–10.

Contact with civilization has both modified the traditional pattern and introduced some new kinds of differentiation. Women nowadays occasionally do work formerly confined to men (e.g., ploughing), and several new occupations (e.g., teaching) are common to both sexes. Owing largely to labour migration, men have on the whole tended more readily than women to discard old tribal practices. But the majority of professing Christians are women, and, in the Protectorate, many more girls than boys have been to school.[27] Daughters now share in the inheritance of cattle from their father's estate,[28] and some women, as teachers, nurses, etc., have acquired an economic independence formerly unknown. In the main, however, the traditional forms of legal and political discrimination still persist almost everywhere.

Age forms the basis for further social distinctions.[29] In family life, people are entitled to respect from those younger than themselves, whose services they can freely command. Children are taught to honour and obey their elders, and may be severely chastised for insolence or undue familiarity. In the relationship terms, and in the associated patterns of behaviour, a man distinguishes clearly between his older and his younger brothers, and he invariably takes precedence over the latter. The intimacies of a common domestic life at times overshadow this hierarchy of age, but on all formal occasions it is emphasized. The regard for one's elders is extended beyond the family and kin to the tribe as a whole. In general, people are expected to respect and obey all those older than themselves, whom they habitually address as " father " (*rrê, ntatê*) or " mother " (*mmê*), and breach of this rule is a penal offence in tribal law.

Children are grouped roughly, according to physical development, into *masea* (sucklings), *banyana* (infants, from about three years old to eight), and, finally, *basimane* (boys) or *basetsana* (girls). Those old enough to be eligible for the next initiation ceremony constitute a special class, known in the Eastern tribes as *magwane*, and in the West by a variety of names (*majafêla, maphatisi*, etc.). The boys of this group, in the East, have costumes, songs, dances, and gatherings peculiar to themselves, and are allowed considerable freedom in conduct, especially in matters of sex.[30] They spend most of their time at the cattle-posts out in the veld, looking after the livestock; the girls help the women at home by fetching water, stamping corn and preparing food, sweeping out the huts, and acting as nursemaids for babies often not much smaller than themselves.

Age-sets

Adults are grouped into formal age-sets or (as they are commonly termed in South Africa) " regiments " (*mephatô*, sing. *mophatô*).[31] There are regiments of men, and regiments of women, and each regiment consists of all tribesmen of the same sex and about the same age. A new regiment is created by the chief every four to seven years or so, when all the eligible boys or girls, aged roughly from 16 to 20, are initiated together; they always include a member of the chief's family, who is henceforth their accepted leader. The creation of a male regiment was formerly accompanied by elaborate ceremonies, known as *bogwêra*, whose most conspicuous features were circumcision and a period of seclusion in the veld, during which the youths were subjected to hardships of various kinds, graded in order of precedence, and taught tribal laws and traditions. There was a corresponding ceremony, known as *bojale*, for girls; it was held at home in the villages, and included dancing and masquerades, some form of physical operation (e.g., branding on the inner thigh), severe punishments and other hardships, and formal instruction in matters concerning domestic and agricultural life, sex, and behaviour towards

[27] See above, p. 17.
[28] See below, p. 43.
[29] Schapera, 1938, pp. 29 f. ; 1940 (a), pp. 111 f., 260 ; Breutz, 1941, pp. 109 f.
[30] Schapera, 1933 (a), pp. 69 f.
[31] Schapera, 1938, pp. 104–17 ; Breutz, 1941, pp. 110–12 ; Language, 1943 (c), pp. 110–34.

men.[32] These ceremonies are still observed in some tribes (e.g., Malete and Tlôkwa of the Protectorate), but owing to the influence of Christianity have long been abandoned by the majority; the regimental system itself, however, persists everywhere, and new regiments continue to be created from time to time.

Each regiment has its own name, given by the chief at the time of initiation and retained ever afterwards,[33] and to address a person by the name of his regiment is considered a form of special politeness and honour. The name is usually derived from some contemporary incident (e.g., *maYakapula*, " those who go with rain," i.e., during whose initiation there was heavy rain, or *maLwêlamotse*, " those who fight for the town," i.e., who were initiated during a period of civil war); but in several Eastern tribes names occur in cyclical sequence (the number, usually fairly large, varying from one tribe to another). Until they belong to a regiment, people are not regarded as adults, and, in the old days, were not permitted to marry. Members of the same regiment not only work and (if men) fight together;[34] they are also intimate companions and equals (*balekane, bankane,* " mates "), and have strong solidarity as a group. They must, on the other hand, respect members of all regiments formed prior to their own, and are, in turn, treated with deference by their juniors. Breaches of discipline in these and other matters associated with the regimental organization are dealt with by special *ad hoc* courts presided over by the regimental leaders.

Domestic Groupings

The Household[35]

This is the smallest well-defined social unit. It consists basically of a man with his wife or wives and their unmarried children, but often includes one or more married sons, brothers, or even daughters, with their respective families. Polygamous households were formerly common; in 1850 Livingstone recorded that, of 278 married men among the Kaa, 121 (43%) were polygamists, of whom 94 had two wives each, 25 three wives, and two four wives.[36] Chiefs and other important men often had many more; the Kgatla chief Kgamanyane (1850–74) is credited with at least 46, the Ngwato chief Sekgoma I (1835–75) with 16, and the Rolong chief Montshiwa (1849–96) with 12. Nowadays, owing to the spread of Christianity and other civilizing influences, polygamy is relatively rare, although it is nowhere officially prohibited (except, by tribal edict, to younger men among the Ngwaketse). In the Tswana districts of the Protectorate, out of 33,260 married men for whom the data were recorded in the 1946 census, all but 3,587 (11%) had only one wife, and of the polygamists 3,161 had two wives each, 350 three wives, 52 four wives, 18 five wives, and six (all but one in the Ngwato Reserve) six wives or more.[37] Modern households are consequently smaller than before; they may contain up to 15 people, occasionally more, but the average is from five to seven.[38]

Every household has its own compound (*lolwapa, lapa*), consisting of one or more huts and granaries situated within a courtyard surrounded by a reed fence

---

[32] Descriptions of the ceremonies are given by Lemue, 1854, pp. 208–13; Mackenzie, 1871, pp. 375–9; Fritsch, 1872, pp. 205–7; Willoughby, 1909, pp. 228–45; 1923, pp. 128–38; Brown, 1921, pp. 419–27; 1926, pp. 73–90; Schapera, 1938, pp. 105–8; 1940 (a), pp. 255–60; Language, 1943 (c), pp. 113–21; Breutz, 1941, pp. 53–63.
[33] Lists of regimental names have been published by Norton, 1922, pp. 245–51 (general); Schapera, 1938, pp. 310–7 (general); Ellenberger, 1937, pp. 65–70 (Malete); 1939, pp. 191–7 (Tlôkwa); Language, 1943 (c), p. 134 (Tlhaping); van Warmelo, 1944 (a), p. 9 f. (Mosêtlha-Kgatla).
[34] See above, pp. 28 f.
[35] Schapera, 1938, pp. 12–15, 150–64, 169–84; 1940 (a), pp. 94–107 and chaps. vi–xii; Scroggie, 1946, pp. 24, 30 f., and *passim*; Breutz, 1941, pp. 26–35 *passim*.
[36] Quoted by J. J. Freeman, *A Tour in South Africa* (1851), p. 280.
[37] *Census Report*, 1946, Table IV (h).
[38] On the social composition and size of the household, cf. the sample censuses given by Schapera, 1935, pp. 208–21 (Kgatla and Ngwato); 1940 (a), pp. 97–101 (Kgatla); 1943 (a), pp. 94–6 (Ngwaketse); 1946, pp. 41–4 (Tlôkwa).

(Tawana), a wooden palisade (Ngwato, Kwena, Ngwaketse), or a low earthen wall (most other tribes). Each married couple has its own hut, sometimes shared with the younger children; adolescent children of both sexes may live together in another hut, or share one with an older female relative, and unmarried adults usually also have separate huts, one for the people of each sex. The huts are used principally as bedrooms and stores, most activities taking place in the open courtyard, except in wet weather.

The members of a household build their own home and produce the great bulk of their food; they have their own fields for cultivation, and usually also some cattle and other livestock.[39] Formerly they also made much of their clothing and domestic utensils, but such goods are nowadays often purchased from traders. Land, livestock, huts, and other property are controlled by the household-head, who allocates them for use among his dependants. He is the legal head of the group, entitled to obedience, service, and respect from his wives and children, and responsible in law for their dealings with outsiders; formerly he also prayed and sacrificed on their behalf to the spirits of his dead ancestors.[40] The household, too, is the group within which children are reared and trained in conduct and methods of work, and the centre of the ceremonies connected with birth, marriage, death, and other ritual occasions. Its self-contained character is evident during the agricultural season, when each household lives in comparative isolation at its fields.

In a polygamous household each wife has her own compound, fields, cattle, and domestic utensils, which after her death are inherited by her own children. The wives normally rank in order of betrothal (not of marriage itself), but a woman married to replace a dead or childless wife[41] takes the rank of her predecessor. The first wife betrothed is senior to the others, and her eldest son is heir to the status and unallocated property of the father.

The Family-group[42]

Several different households, living side by side and acknowledging a common " elder " (*mogolwane*), constitute a family-group. Such a group usually contains from 20 to 50 persons. It consists basically of families whose men are all descended agnatically from a common grandfather or great-grandfather, by whose name the group is commonly known; the " elder " is always the man senior to the rest in line of descent. The group may, however, also include relatives of other categories, such as affines and uterine nephews, and perhaps even one or more families of unrelated dependants.[43]

The family-group is the setting for the more important domestic events and activities of its component households. Its members associate together constantly, co-operate in such major tasks as building and thatching huts, clearing new fields, weeding, and reaping, and help one another with gifts or loans of food, livestock, and other commodities. It deals as a unit with such matters as betrothal and marriage negotiations, the organization of feasts, the settlement of estates, and the future of widows, all of which are held to concern not one household alone but the group as a whole. Its men also meet, under the leadership of the " elder," to arbitrate over internal disputes, and only if an acceptable compromise cannot be reached, or if the issues involved are fairly serious, is the matter then referred to the official tribal courts; moreover, should any member be involved at law with outsiders, his " elder " and the other men are expected to support him at any preliminary discussions and in the formal hearings at court.

[39] See above, p. 27.
[40] See below, p. 59.
[41] See below, p. 42.
[42] Schapera, 1938, pp. 15 f., 89–91 ; Breutz, 1941, pp. 26–35.
[43] For the size and composition of typical family-groups, cf. Schapera, 1946, pp. 30 f. ; 1952 (b), pp. 49 f., 90, 91, 102.

Marriage[44]

A man's first wife is generally selected for him by his parents; any others he chooses himself. Normally he may not marry such close relatives as his sister, parent's sister, or sibling's child, but some tribes (e.g., Ngwaketse and Rolong) formerly permitted marriage with an agnatic half-sister, and many others still allow it with a brother's daughter, although not with a sister's. A few (e.g., Kgatla, Tlôkwa, and Ngwaketse) are exceptional in also prohibiting marriage with the daughter of a regimental age-mate, and most Tswana in the Protectorate will not marry women of such servile stocks as the Sarwa.[45]

All Tswana, on the other hand, encourage marriage with first cousins. Preference is usually expressed for a cross-cousin, especially on the mother's side; a father's brother's daughter is also favoured, but some tribes are doubtful about the suitability of a mother's sister's daughter, although she is nowhere explicitly prohibited.[46] Failing a cousin, any more remote relative may be chosen. Most men, however, marry women to whom they are not related at all. In the Protectorate, of the marriages contracted by a sample group of 1,599 commoners (from five different tribes), 20.7% were between relatives, 5.5% being between people with a common grandparent or more closely related; the corresponding proportions for a group of 1,546 nobles (from eight different tribes) were much higher, being 50.8% and 11.4% respectively. The cousin most frequently married by commoners was the mother's brother's daughter, but by nobles the father's brother's daughter, although in both groups there were also marriages with every other type of first cousin.[47]

Marriages are usually arranged by negotiation between the family-groups concerned, the boy's people taking the initiative. In the old days child betrothals were fairly common, although the wedding was delayed until the boy and girl had both been initiated into their regiments. A feast was then held at the girl's home, after which the boy came there in the evenings and stayed with her in a hut specially set aside for them, returning by day to his own parents' home. This practice, known as *go ralala*, was usually continued until a child had been born. The woman was then taken formally to live at her husband's home, the occasion being marked by special festivities. The *ralala* custom has now been abandoned in some tribes (e.g., Ngwato and Tawana), but is still found in others (e.g., Ngwaketse and Kwena), although even here it is not universally observed. Child betrothal is no longer practised, and it is common for people to choose their own mates, the boy's family then approaching the girl's on his behalf. The usual age of marriage varies from about 21 to 25 for women, and 25 to 30 for men. The traditional forms of wedding feast are still widely observed, even in the " more respectable " instances of marriage by Christian or civil rites. In the Protectorate, according to the 1946 census, about 12% of all marriages in the Tswana districts have been contracted by such rites.[48]

Mutual agreement between the family-groups concerned, as reflected in the formalities of betrothal, is an essential condition of marriage. Another, formerly, was the transfer of livestock from the bridegroom's family to the bride's. These animals, known as *bogadi*, generally consisted of large cattle only, but in some tribes (e.g., Ngwaketse and Kwena) sheep were also given, but never goats. The number was decided by the groom's people alone, the bride's having no say in the matter; the average varied from tribe to tribe, but was usually from four to ten head.

[44] Schapera, 1938, pp. 125–47 ; 1940 (a), pp. 38–92 ; Breutz, 1941, pp. 35–46 ; Brown, 1926, pp. 58–65 ; Willoughby, 1923, pp. 108–18 ; Matthews, 1940, pp 1–23 ; Ellenberger 1937, pp. 63 f. ; Whitfield, 1948, pp. 210–2 ; Language, 1943 (a), pp. 69–80.
[45] Schapera, 1949, pp. 104–20 ; Matthews, 1940, pp. 9 f.
[46] But Language says that among the Tlhaping marriage was in the old days prohibited between a man and the daughter of his mother's sister or of his father's full brother (1943 (a), p. 71)
[47] Schapera, 1950, pp. 149–52, 154–8.
[48] *Census Report*, 1946, Table IV (g).

The animals were contributed by the groom's father and other relatives, especially his maternal uncle, and after delivery were distributed among the wife's kin, her " linked " maternal uncle and brother[49] being among those with preferential claims. They were normally due when the wife went to live at her husband's home, but were often delayed for many years. In Tswana law, however, no marriage was considered valid unless and until *bogadi* had been given. *Bogadi* has now been abolished among the Ngwato and Tawana, and has been widely abandoned by Christians in most other tribes; it is still given by other people, however, and in the main follows the pattern described.[50]

Certain other usages associated with marriage have also been widely abandoned. In the old days, a man whose wife was barren or had died childless might claim another woman from her family to replace her; the children borne by this " substitute wife " (*seantlo*) were held to belong to the " house " of her predecessor. Similarly, a widow if young enough was taken over by her late husband's younger brother or some other close agnatic kinsman, the children he begot by her ranking as the legal offspring of the deceased. This custom is still practised occasionally, although as a rule widows are now free to marry outside their late husband's family; on the other hand, the provision of a *seantlo* is virtually extinct.[51]

Divorce[52]

In the Protectorate, at the time of the 1946 census, 2.3% of the married men, and 4.9% of the married women, were described as " divorced "; the figures do not show how many people had married again after a divorce. Data previously collected suggest that relatively few marriages, certainly well under 10%, end in divorce.[53] When husband and wife quarrel seriously, their relatives intervene, and only if no reconciliation can be effected is the matter taken to court. Marriages contracted by civil or Christian rites cannot be dissolved except by the Government courts; all others are dealt with by the chief, who also tries first to reconcile the couple. In tribal law the main grounds for divorce are a husband's sorcery, cruelty, or non-support, and a wife's sorcery, barrenness, desertion, or failure to perform her domestic duties. Adultery, a valid ground in civil law, is seldom treated as such in the tribal courts, unless often repeated; usually the husband merely claims compensation from his wife's lover, whereas if he himself is the offender she has no legal remedy. A divorced woman returns to her parental home, and can then be married again. She is usually given some corn and other household property, and, if the injured party, some cattle as well. Her children, if still young, accompany her, but when old enough return to their father, who in the meantime contributes to their support. He is never entitled to the return of his *bogadi;* among the Kgatla and some other tribes he could formerly reclaim it, if the marriage was childless, but even this right has now been abolished.

Succession and Inheritance[54]

A man's eldest son normally succeeds him as head of his household and to any political office that he may have held, and also inherits the great bulk of his cattle and such other property (e.g., money, ploughs, and guns) as remains unallotted. The younger sons are likewise given a few cattle each, although usually their father will have provided for them during his lifetime. The widow and daughters formerly

---

[49] See below, p. 45.
[50] Lestrade, 1926, pp. 937–42 ; Breutz, 1941, pp. 40, 43 ; Matthews, 1940, pp. 13–19 ; Willoughby, 1923, pp. 108–17 ; Schapera, 1938, pp. 138–47 ; 1940 (a), pp. 82–91 ; 1943 (e), pp. 40 f. ; Jennings, 1933 (a missionary attack upon *bogadi*).
[51] Schapera, 1938, pp. 155 f., 164–8 ; 1940 (a), pp. 214 f., 316–25 ; 1950, pp. 152–4, 158–9 ; Lestrade, 1926, pp. 941–2 ; Language, 1943 (a), pp. 84–88.
[52] Schapera, 1938, pp. 159–63 ; 1940 (a), pp. 288–303.
[53] *Census Report*, 1946, Table IV (g) ; Schapera, 1940 (a), p. 294.
[54] Schapera, 1938, pp. 53–7, 230–8 ; 1940 (a), pp. 327–32 ; 1943 (a), pp. 87–9, 152–5 ; Language, 1943 (a), pp. 80–4, 88–91 ; Whitfield, 1948, pp. 210–2, 366–72.

received no cattle at all, but nowadays almost everywhere share in the estate, although to a much smaller extent than the sons. The huts, fields, etc., are retained by the widow as long as she lives; after her death, the huts are inherited by the youngest surviving son, the fields are distributed among adult children (both male and female) for whom provision has not yet been made, and the household utensils go to the daughters, the eldest receiving the most. Clothes and certain other personal belongings are the perquisite of the dead man's linked maternal uncle, who also receives a bull from his herd.

In a polygamous household, the eldest son of the " great " wife is principal heir both to the general estate of his father and to the property of his own mother's " house." The eldest son of every other wife is the principal heir in her " house," and becomes the responsible guardian of his full siblings, but in matters affecting the household generally he remains subordinate to the " great son " (*mojaboswa*, " the eater of the inheritance ").

If there are no sons at all, or if they are still minors, the dead man's estate comes under the control of his nearest male agnate, usually his next younger brother. This man supports the widow and children from the property entrusted to him, and is legally responsible for looking after their welfare. Formerly, in the absence of sons, he himself inherited the estate after the widow's death and the marriage of her daughters, but nowadays the daughters are usually given preference; he still succeeds, however, to any office that the deceased may have held.

KINSHIP[55]

Tswana carry recognition of kinship much further than is common in Western European society. A man differentiates between his close relatives according to sex, relative age, and line of descent, and applies a special kinship term to each category. These terms he extends in various ways to more distant relatives, so that almost everybody with whom he can establish genealogical connection, no matter how remote, is brought within his circle of kin. These people, unlike the members of his family-group and ward, are widely dispersed throughout the tribe, and possibly even beyond; on the other hand, especially in the larger tribes, most of his fellow-tribesmen are not related to him at all.

The following is a list of the terms chiefly used for closer kinsmen and affines, arranged in pairs of reciprocals. Where more than one term may be used, the commoner is mentioned first; an asterisk preceding a term denotes a dialectal variation.[56] (All terms are given here in their simplest form; various modifications can be made to indicate person or number.)

Parent : child
    1. (*a*)   F = *rrê*, *ntatê ; m = *mmê*
       (*b*)   c = *ngwana* ; S = *morwa* ; d = *morwadi*

Siblings :
    1. (*a*)   OB (m.s.), osi. (w.s.) = *mogolole*, *nkgonne
       (*b*)   YB (m.s.), ysi (w.s.) = *nnake*
    2. (*a*) B (w.s.), (*b*) si. (m.s.) = *kgaitsadi*, *kgantsadi

Grandparents : grandchildren :
    1. (*a*)   FF, mF = *rrêmogolo*, *ntatêmogolo
            Fm, mm = *mmêmogolo*, *nkoko
       (*b*)   Sc, dc = *ngwana-ngwana*, *setlogolo, *motlogolo

---

[55] Schapera, 1938, pp. 17–19, 185–90, etc. ; 1940 (a), pp. 108–15 ; 1950 pp. 142–9 ; Breutz, 1941, pp. 26–35 ; Brown, 1926, pp. 52–7 ; Scroggie, 1946, pp. 24–31 ; van Warmelo, 1931, pp. 60–63, 72–87.

[56] Abbreviations used are : F = father, m = mother, c = child, S = son, d = daughter, B = brother, si = sister, H = husband, w = wife, P = parents ; OB, osi = elder brother, elder sister ; YB, ysi = younger brother, younger sister ; m.s. = man speaking, w.s. = woman speaking ; /, as in *ngwana-mogolole/nnake* = *ngwana-mogolole* or *ngwana-nnake*.

Parent's sibling : sibling's child :
1. (a) F.OB = rrêmogolo, *ntatêmogolo
   F.YB = rrangwane
   (b) B.c. (m.s.) = ngwana-mogolole/nnake ; *ngwana
2. (a) F.si. = rrakgadi
   (b) B.c. (w.s.) = ngwana-kgaitsadi
3. (a) m.B. = malome
   (b) si.c. (m.s.) = setlogolo, *motlogolo
4. (a) m.osi. = mmêmogolo, *nkoko
   m.ysi. = mmangwane, *mmane
   (b) si.c. (w.s.) = ngwana-mogolole/nnake ; *ngwana

Cousins :
1. F.B.c. = sibling ; ngwana-rrêmogolo/rrangwane
2. m.si.c. = ngwana-mmêmogolo/mmangwane ; sibling ; m.si.
3. (a) m.B.c., (b) F.si.c. = ntsala, *motswala

Spouses :
1. (a) H = monna (" man "), mogatsa
   (b) w = mosadi (" woman "), mogatsa

Wife's parent : daughter's husband :
1. (a) w.F., w.m. = mogwagadi
   (b) d.H. = mogwê, *mokgonyana

Wife's sibling : sister's husband :
1. (a) w.B. = B. (m.s.) ; w.F. ; *mogwê, *molamo
   (b) si.H. (m.s.) = d.H. ; B. (m.s.) ; *molamo
2. (a) w.si. = si. (w.s.) ; *w.m.
   (b) si.H. (w.s.) = si. (w.s.) ; d.H. (ysi.H. only)

Wife's sibling's child : aunt's husband :
1. (a) w.B.c. = B.c. (w.s.) ; B.c. (m.s.) ; w.P.
   (b) F.si.H. = F.si. ; *d.H.
2. (a) w.si.c. = si.c. (w.s.) ; *ngwana-mogwagadi
   (b) m.si.H. = m.si. ; F.B. ; mogatsa-mmêmogolo/mmangwane

Husband's parents : son's wife :
1. (a) H.F. = matsale, *rratswale
   H.m. = matsale
   (b) S.w. = ngwetsi

Husband's sibling : brother's wife :
1. (a) H.B. = B. (m.s.) ; H.F. (H.OB. only)
   (b) B.w. (m.s.) = B. (m.s.) ; S.w. (YB.w. only)
2. (a) H.si. = si. (w.s.) ; *mogadibo, *mogokane, *bofije
   (b) B.w. (w.s.) = H.si.

Husband's sibling's child : uncle's wife :
1. (a) H.B.c. = B.c. (m.s.)
   (b) F.B.w. = m.si.
2. (a) H.si.c. = si.c. (m.s.)
   (b) m.B.w. = m.osi. ; mogatsa-malome ; m.B.

Spouse's sibling's spouse :
1. (a) w.B.w. = w.m. ; si. (w.s.)
   (b) H.si.H. = B. (m.s.) ; d.H.
2. w.si.H. = B. (m.s.) ; *kupa, *foba, *šobe
3. H.B.w. = si. (w.s.) ; *molodikane

Child's spouse's parents :
1. (a) S.w.P. = w.P.
   (b) d.H.P. = d.H. ; *H.P.

Kinship forms the basis of many social institutions. The emphasis is predominantly patrilineal. Membership of tribe, ward, etc., is determined primarily by descent traced through the father; property and rank normally pass from father to son, or, failing a son, to the next male member of the same lineage (*lesika*); a man's surname is usually the given name of his father or paternal grandfather, and wards or family-groups often bear the name of their leader's agnatic ancestor; and, in the old days, people worshipped the ancestors of their deceased paternal ancestors. As already mentioned, too, kinship is the main determinant of marriage prohibitions and preferences.

Relatives of all kinds are expected to be friendly and hospitable, and to help one another at work, with gifts of food, clothing, etc., and in times of trouble. These obligations apply especially to close relatives, e.g., a man's own siblings, his parents' siblings and their families, or his wife's parents and siblings. They advise and help him in all his troubles and undertakings, a special " family council " (*phuthêgô ya bana ba mpa*) being summoned if necessary; they also exchange with him specified gifts and ceremonial services, whose nature varies with the type of relationship. His more distant relatives, unless they are also near neighbours, seldom figure prominently in his everyday life, but are invited to attend on all occasions of domestic importance.

The significance of close kin is reflected in the custom of *go rulaganya ga bana*, to " link " siblings together; their father assigns brother to brother, sister to sister, and sister to brother, so that every child has a linked sibling of each sex. A man's linked sister in due course becomes the " special " paternal aunt of his children, and he becomes the special maternal uncle of hers; similarly, linked brothers (or sisters) become the special senior or junior paternal uncles (or maternal aunts) of each other's children. Among his close relatives, consequently, he has one of each category to whom he is specially attached, and with whom he is said " to work together for life "; it is they, above all, who carry out the obligations conventionally assigned to relatives of their class.

A man's close agnatic kin (*ba ga etsho*) usually belong to his own family-group or ward; consequently they are also the relatives with whom he habitually associates, and from whom he expects immediate support and protection. But personal relations are also governed by well-defined principles of discipline and authority. In general, a man is entitled to obedience and respect from those junior to him in line of descent; he, in turn, must defer to all his seniors. Normally the system works reasonably well and without overt friction, but disputes sometimes arise owing to arbitrary exercise of authority and rival claims to property or position, a common result being that the group breaks apart.

A man's maternal relatives (*ba ga etsho mogolo*) generally belong to some other ward, and, owing to the rules of inheritance and succession, cannot be his rivals for property or position. They are notoriously more attached and devoted to him than are his agnates. Children are often sent on long visits to their mother's people, and while there enjoy many privileges. A man looks above all to his linked maternal uncle for disinterested advice and aid in times of difficulty; the uncle must also be consulted in all matters specially affecting his sister's children, and has a decisive say in the arrangement of their marriages. Between his own children and his sister's there exists an institutionalized and symmetrical joking relationship (*go tlhagana*), most commonly expressed in exchanges of ribaldry; as already noted, too, a cross-cousin is considered the most suitable person for one to marry.

Affines, whether or not also related by blood, share in the general pattern of friendship and co-operation. While cohabiting with his wife under the *ralala* custom[57] a man sees much of her parents and other close relatives, who, without being familiar, treat him cordially and with respect. After taking his wife to his own home he frequently visits his relatives-in-law (*bagwagadi*), helps them at work, and consults them in matters of domestic concern. They reciprocate in the same manner. A woman's relations with her husband's people (*bo-matsale*) may at first

---

[57] See above, p. 41.

be characterized by mutual politeness, accompanied perhaps by a show of authority on their part; but if she proves a good wife they gradually unbend and she is accepted and treated as a daughter. There is no form of taboo or other prescribed avoidance between relatives-in-law of any kind.

WARDS AND SECTIONS[58]

Most family-groups are subdivisions of wards. The ward (*motse, kgotla, kgôrô*) is a patrilineal but non-exogamous body of people forming a distinct social and administrative unit under the leadership and authority of a hereditary headman. Wards vary considerably in size; some have less than 100 people each, and others well over 1,000, but the majority contain between 300 and 600. The members consist, firstly, of the headman's close relatives; some of the smallest wards are in fact single family-groups. But the great majority also contain other family-groups, sometimes remoter segments of the headman's lineage, but often not related either to him or to one another, except perhaps through intermarriage. Such groups, sometimes sufficiently large and autonomous to be classed as sub-wards (*makgotlana*), may have been included in the ward when it was founded, but mostly are later accessions. A man normally belongs for life to his natal ward, but may be transferred to another by the chief (e.g., when a new ward is created, or in case of serious internal dispute).

Each ward has its own name, its own settlement or settlements, and one or more cultivating areas of its own. Its members associate together habitually, co-operate in many activities, and support one another in case of trouble with outsiders. A man refers to his fellow-members as *ba ga etsho*, " the people of my home " (which is also what he calls his near agnatic kin), and sometimes applies to them all, whether actually related to him or not, the kinship terms that he uses for his close agnates. The ward is also the basic unit in the administrative system: it has its own court, where the headman judges complaints lodged against any of his people;[59] its members, headed in each case by the most senior in rank, constitute separate units in the age-sets; they may be required, as a body, to do some form of tribal work; and they are sometimes grouped apart at feasts and political assemblies. Certain wards are also attached as personal retainers to the household of the chief.[60]

The number of wards in a tribe varies roughly with its size; thus, in the Protectorate, the Ngwato have 310, the Ngwaketse 119, the Kwena 69, Kgatla 68, Tawana 40, Malete 12, and Tlôkwa 5.[61] But the number is not constant. New wards are created from time to time, e.g., to serve as personal retainers of the chief's principal sons, or to look after some of the royal cattle; a group of immigrants, if large enough, will be classed as a separate ward, with its leader as headman; and an existing ward may divide, the offshoot then being recognized as independent. Only the chief can create or recognize a new ward. Wards may also disappear, e.g., if they secede from the tribe, or if they become so small that the chief decides to amalgamate the remnant with some other ward. The chief himself is head of the royal ward, and the other ward-heads usually include nobles, commoners, and immigrants.[62]

All members of a ward are expected to live together. This rule does not apply to subject peoples of alien culture (as in the northern Protectorate), but is still commonly observed by the Tswana themselves, except in parts of British Bechuanaland. A large proportion of the wards are located in the capital, where each has its own hamlet; some of the other big villages also contain several wards, but each of

---

[58] Campbell, 1813, p. 255; 1820, Vol. II, pp. 152–4; Schapera, 1935, pp. 203–6, 221–4; 1938, pp. 19–28; 1940 (b) pp. 58–62; 1943 (a), pp. 28–30, 44–6, 68–77, 143–8; 1946, pp. 22–31; Ashton, 1937, pp. 75 f., 80–2; Ellenberger, 1937, pp. 55–8; Language, 1943 (a), pp. 8–15; Scroggie, 1946, pp. 17–19.
[59] See below, p. 55.
[60] See above, p. 28.
[61] The names and histories of these wards are given in Schapera, 1952 (b), pp. 39–124.
[62] For the rare exception to this rule, cf. Schapera, 1952 (b), pp. 67, 116.

the smaller villages is usually inhabited by one only. The ward settlement, whether isolated or part of a larger cluster, is typically circular in form. The dwellings are built close together round the circumference, leaving a large open space in the centre. This central space usually contains some trees, to provide shelter from the sun, and one or more kraals for livestock temporarily brought in from the cattle-posts. Adjoining the kraals is a windbreak of stout poles defining the council-place (*kgotla*), which is the focus of the ward's public life; here cases are tried and official meetings held, and here the men habitually forgather round a fire in the early morning and late afternoon. In a large ward there are several adjoining settlements of this kind, each inhabited by a sub-ward or family-group.

In the bigger villages, especially the capitals, the ward hamlets are usually situated according to traditional rules of seniority and precedence, and in the old days, whenever the village moved, the pattern was repeated in the new settlement. In most tribes, three local divisions are recognized: central (*fa gare*), upper (*ntlha ya godimo*), and lower (*ntlha ya tlase*). The names are said to be derived from the days when villages were habitually built on the banks of rivers, *godimo* being "up stream" and *tlase* "down stream," with the chief's division in the centre. The divisions were formerly of little social importance, except that they ranked in the order given above, and that all the wards in each constituted a single section of the tribal army.[63] Since 1938, however, they have been used in the Protectorate (among the Kwena and Ngwaketse) as the basis of reforms in the judicial organization; the courts of the individual ward-heads in each have now been superseded by a single court, with the senior noble ward-head of the division as judge.

The Ngwato, Tawana, and Kgatla have a different system of organization.[64] All the wards in the tribe (and not merely those of the capital) are grouped into "sections" (also termed *kgotla* or *kgôrô*); the Ngwato have four, Tawana three, and Kgatla five. The sections in each have been developed over the course of time, being stabilized about the middle of the 19th century, and differ in detail from one tribe to another. Among the Kgatla they are ranked in a well-defined order of precedence; the first includes almost all the wards under noble and immigrant headmen, and the others are composed mainly of wards under commoners. In the two other tribes, each section contains wards under nobles, commoners, and immigrants; they are not differentiated in rank among the Ngwato, but among the Tawana one of them is headed by the chief, and therefore takes precedence over the others. In all three tribes, one of the ward-heads in each section is senior to the rest, and exercises an overriding authority in judicial and administrative life; the section often meets alone under his leadership to discuss matters of local or tribal concern; its members constitute a separate unit in each age-set; and, in the capital, all the local wards belonging to each live side by side. A similar system of grouping may possibly exist elsewhere (e.g., among the Tshidi-Rolong and the Mosêtlha-Kgatla), but the details recorded are too scanty to give any clear indication.

Modern Groupings[65]

Since the impact of Western civilization, several new forms of social grouping have developed. By far the most widespread are the Christian churches. In virtually every tribe one or more missionary societies are at work, and many of the people have become professing Christians. In the Protectorate, they constitute 23% of the population in the Tswana districts, although the proportion varies greatly from one tribe to another (e.g., Kgatla 65%, Ngwaketse and Kwena 38%, Tawana 16%, and Ngwato only 7%);[66] separate figures are not available for Tswana in the Union, but 51% of the total Native population there are described as "members of Christian

---

[63] Schapera, 1943 (a), pp. 69–71; 1946, pp. 24 f.
[64] Schapera, 1938, pp. 24–8, 101–3.
[65] Schapera, 1934 (c), pp. 50–7; 1936, pp. 232–5; Scroggie, 1946, pp. 22 f.
[66] *Census Report*, 1946, Table IV (i).

churches," and another 8.8% as belonging to Native separatist churches.[67] As a group, Christians are distinguished not only by their system of worship, with its own rituals and beliefs; they must also conform to the social and moral ideals preached by the Church, dress in a " respectable " manner, and abstain from certain tribal customs regarded as incompatible with true Christianity (e.g., polygamy, circumcision, and often also *bogadi*). They do not always live up to these standards, which nevertheless constitute a recognized line of demarcation between them and other people. Often two or more rival denominations are represented in the tribe, with the result that the Christians are themselves divided into different groups.[68]

Education has created another line of division.[69] As already mentioned, schools have been built in virtually every tribal area, and about one-fifth of the Tswana in the Protectorate are at least able to read.[70] They are nowhere organized as a single group, except while at school, but the influence of education is reflected in the widespread existence of teachers' associations, reading and debating societies, such youth movements as the Pathfinders (for boys) and Wayfarers (for girls), and such sports associations as football and basketball clubs, which sometimes compete on an intertribal basis. There is often also an appreciable difference in standards of living and outlook between the more highly educated people and the rest of the tribe.

Both teachers and local Church leaders have special authority within their respective fields. But apart perhaps from them, and the magicians, there does not yet exist what may be termed a professional class. Many other new forms of occupation have indeed developed. The Government and tribal administrations employ clerks, tax-collectors, policemen, agricultural demonstrators, nurses, etc., local European residents have shop hands and domestic servants, and some people work for themselves as traders, butchers, dressmakers, carpenters, thatchers, etc. But these developments have not, in general, led to concentration on one particular calling to the exclusion of all others; almost every tribesman remains primarily a herdsman and cultivator, and the specialist calling in which he may also be engaged is still supplementary to his farming activities, and has by no means taken their place. All this, nevertheless, has led to much greater variation in custom than under the old tribal system. Some people are very conservative, whereas others have discarded many traditional usages and adopted European ways of life. Thus, among the Mosêtlha-Kgatla of the Transvaal, there existed in 1941 an " Improvement Society " of five years' standing, consisting of " a group of go-ahead men and women, with a uniform, a song, and a programme of reform."[71]

Some of the Tswana working away from home, especially on the Witwatersrand, have recently also tended to group together, usually on a tribal basis, into mutual benefit associations. The Kgatla, for instance, have had one of this kind since 1933; its functions include helping unemployed fellow-tribesmen and repatriating those who are sick or unable to pay the fare home.[72]

---

[67] *Report of Commission on Native Education* (U.G.53, 1951), p. 12.
[68] See below, p. 58.
[69] Schapera, 1940 (a), pp. 262–4, 267–8, 337–8.
[70] See above, p. 18.
[71] Richards, 1941, p. 90.
[72] Schapera, 1947 (a), p. 168.

GOVERNMENT AND LAW

European Control

Central Government

All Tswana tribes are nowadays under European sovereignty and subject to laws imposed from outside. In the Union, legislative power is vested in a Parliament consisting of two houses, the Senate (with 44 members) and the Assembly (159 members). Four senators and three members of the Assembly, all of whom must be Europeans, are elected on a separate roll by the Native population; British Bechuanaland participates with other districts of the Cape Province in electing one member for each house, and another senator is elected for the whole of the Transvaal. This is the only direct representation that the Tswana have in Parliament. The Native Administration Act (No. 38 of 1927) constituted the Governor-General of the Union "supreme chief" of all Natives, except in the Cape Province, and authorized him to legislate by Proclamation for Native areas, both generally and individually; all Proclamations, however, must receive Parliamentary approval. Legislation and all other Government matters specially affecting Natives are administered by the Department of Native Affairs, which is controlled by a special Cabinet Minister and run by a permanent Secretary with many assistants. The Minister is advised by a statutory Native Affairs Commission, of not less than three or more than five members appointed by the Governor-General; the Commission considers and makes recommendations on legislation, administration, and general questions of Native policy, including the acquisition of land under the provisions of the Native Trust and Land Act (No. 18 of 1936).[1]

In the Protectorate, legislative and administrative authority is vested in a High Commissioner (stationed in the capital of the Union), who is also in charge of Basutoland and Swaziland; he, in turn, is responsible to the Secretary of State for Commonwealth Relations in Great Britain, who must account to the British Parliament for all decisions on major matters of policy. The Territory itself is under the immediate control of a Resident Commissioner, with headquarters at Mafeking (in British Bechuanaland), who is authorized to act for the High Commissioner in certain matters and to make minor regulations. The local population, whether European or Native, has no direct control over legislation, which is drafted by the High Commissioner and issued by Proclamation; nor, unlike the Union, is there any Government department specially concerned with all Native affairs.[2]

In both Territories, some laws apply to both Natives and other people, e.g., the provisions of the criminal code, and the law relating to civil marriages. Others affect Natives only, or discriminate between them and Europeans. Thus, in both Territories, they may not be supplied with intoxicating liquor other than Kafir beer; every adult male must also pay an annual poll tax, failing which he is liable to prosecution. In the Union, but not in the Protectorate, Natives may not buy or hire land from Europeans outside certain areas scheduled for Native occupation, and those living or working outside tribal areas must carry special passes. There are many other legal discriminations of a similar character.[3]

Native Councils

Although lacking central legislative powers, the Tswana (and other Natives) do participate to some extent in determining Government policy about their affairs. In the Union a Natives' Representative Council was established in 1936. It consisted of

---

[1] *Official Year Book of the Union*, No. 25 (1949), pp. 465-8; Rogers and Linington, 1949, pp. 15-33; Brookes, in Hellmann and Abrahams, 1949, chap. iii.
[2] Ashton, 1949, pp. 707-9, 711; *B.P. Annual Report*, 1950, pp. 42-4.
[3] Rogers and Linington, 1949, chap. viii; Hellmann and Abrahams, 1949, chaps. iv, xi, xii, etc.

the Secretary for Native Affairs as chairman, the six Chief Native Commissioners as assessors, and 16 Native members (four nominated by the Governor-General, and 12 elected by the people); each of the four electoral areas for the Senate was represented here by one nominated and three elected members. In 1949 three of the African members were of Tswana extraction. The Council met at least once a year, and might be summoned more often by the Minister. It discussed the estimates of the South African Native Trust Fund, and all legislation certified as specially or differentially affecting Natives. It also had very full powers of initiative, " and may and does discuss other matters affecting Africans and even general issues of national life. It is, however, a purely advisory body. Its resolutions have no force of law, and are frequently not accepted by the Government." But " there is no doubt at all that it has exercised quite a considerable influence on the deliberations of Parliament and the decisions of the Government."[4] It was, however, abolished in 1951.

In the Protectorate there has existed since 1920 an African Advisory Council, which meets annually under the presidency of the Resident Commissioner. Unlike the parallel European Advisory Council, it is not a statutory body, but it is officially recognized, and its present constitution was approved by the High Commissioner in 1944. It consists of 36 African members, each tribe being represented by its chief and two to seven other men chosen by the people in *kgotla* (i.e., at a tribal assembly); heads of Government departments and District Officers also attend, but do not vote, nor do they speak unless their work is under review. The Council discusses and may pass resolutions on any matter raised either by its African members or by the Administration. Its opinions are treated with respect, and contribute much to the shaping of policy. Within recent years, especially, legislation affecting African interests has always been referred to it before promulgation, and the draft has sometimes been amended in view of the comments made. In 1950 a Joint Advisory Council was also established, consisting of eight members nominated by the African Council, the eight elected members of the European Council, and four official members nominated by the Resident Commissioner. This new Council, which has already met several times, is designed to " enable the representatives of both sections of the population to discuss together their common problems, and to appreciate each other's difficulties."[5]

Local Administration

The Native population of the Union is divided for purposes of local administration into six regional groups, each under a Chief Native Commissioner. Almost all the rural Tswana, in both British Bechuanaland and the Transvaal, fall under the jurisdiction of the C.N.C., Western Areas, whose headquarters are at Potchefstroom (Transvaal). Under him are Native Commissioners in charge of predominantly Native areas; districts with mixed population are administered by Magistrates appointed by the Department of Justice, but there are usually also Additional or Assistant Native Commissioners (appointed by the Department of Native Affairs), who deal with matters affecting Natives only. The principal duties of the Native Commissioner are to administer all laws and regulations applying to Natives, collect taxes, and exercise such civil and criminal jurisdiction as may be conferred upon him; to communicate new laws and instructions to the Natives, inquire into their complaints, and transmit to the Department any representations they wish to make on matters concerning them; to keep informed of all local developments, and help to promote social and economic welfare; and, in general, to watch over the interests of the people.[6] He also presides over meetings of the Local Council, a body, usually representing several different tribes, whose primary concern is to raise and spend

[4] Brookes, in Hellmann and Abrahams, 1949, p. 31 ; cf. Rogers and Linington, 1949, pp. 27 f. ; *Official Year Book*, 1949, pp. 466 f.
[5] *B.P. Annual Report*, 1950, p. 2 ; cf. Ashton, 1947, p. 240.
[6] Rogers and Linington, 1949, p. 7–10 ; *Official Year Book*, 1949, pp. 468–70.

money on social and economic development; as already noted, five such Councils have been established in the Tswana areas.[7]

The Protectorate is divided into 12 districts, several of which consist of single Reserves. Each is administered by a District Commissioner, whose authority extends over both Europeans and Africans. He conveys all Government communications to the tribal authorities, receives and disburses Government moneys, attends tribal meetings to explain new or projected measures, hears appeals from the tribal courts and tries cases excluded from their jurisdiction, supervises the work of the Treasuries and other tribal committees, advises the chiefs and tries through them to promote social and economic development, tours the district regularly to keep in touch with local affairs, and deals with inquiries, grievances, and complaints. As in the Union, he is assisted by a small police force and by subordinate officers of various kinds, mostly Africans; the technical branches of the Administration (e.g., medicine and agriculture) usually also have local staffs of their own, who work under his general direction.[8]

Tribal Administration

Despite their loss of political independence, and the various other changes noted above, the Tswana are still largely self-governing. Subject to the overriding control of the European Administrations, and within the limits imposed by them, each tribe continues to manage its own affairs, and the traditional agents of government are employed almost everywhere. In the Union, tribal self-government is regulated mainly by the provisions of the Native Administration Act (No. 38 of 1927, as amended); in the Protectorate, the corresponding ordinance is the Native Administration Proclamation (No. 32 of 1943), which replaced and repealed an earlier Proclamation of the same title (No. 74 of 1934).

Chiefship[9]

Each tribe has a chief (*kgosi, morêna*), who is normally the lifelong head of its government. In the old days a newly deceased chief was usually succeeded by the eldest son of his great wife, or, if there was no male issue in her house, by the eldest son of the wife next in rank; if there were no sons or male descendants of sons by any wife, the succession passed to the line of the chief's next younger brother, and so on. The order of succession was determined primarily by seniority of descent; sometimes, however, doubts about the relative status of wives led to disputes between rival claimants, followed possibly by a split in the tribe. Nowadays the chiefship is still hereditary according to the traditional rules of precedence. But, in both the Union and the Protectorate, succession to the office depends also upon the approval of the Government, which may reject the heir apparent if he is considered unsuitable, and appoint a junior kinsman instead; moreover, unless and until explicitly recognized or appointed by the Government, no chief is entitled to rule over his tribe. If an heir is too young to rule when his father dies, a paternal uncle or some other close agnate is chosen to act as chief during his minority.

The chief's duties formerly extended over many different spheres of tribal life. With the aid of his advisers and councils he decided questions of public policy, promulgated new laws and amended or abolished others, organized regimental and other large-scale activities, and took whatever action seemed appropriate in case of war, pestilence, famine, or some other calamity. He shared his wealth with his subjects, and his personal popularity often rested on his reputation for generosity. He judged all serious crimes and civil disputes, heard appeals from the verdicts of lesser courts, controlled the activities of ward-heads and other local authorities, and regulated the distribution and use of land, the cycle of agricultural work, and many

---

[7] See above, p. 33.
[8] Ashton, 1947, pp. 239 f.; 1949, pp. 711 f.; Pim, 1933, pp. 55–63.
[9] Schapera, 1938, pp. 53–72, 84–8; 1940 (b), pp. 66–71, 74–82; Ashton, 1947, pp. 235–9, 242–8; Language, 1941, pp. 121–99, 264–360; Breutz, 1941, pp. 74–95; Hailey, 1938, pp. 402–13.

other economic matters. He often led the tribal army in war, and arranged or personally performed many religious and magical ceremonies on behalf of the tribe.[10] He was expected to watch over the interests of his subjects, and keep informed of tribal affairs generally; he therefore spent much time daily at his *kgotla* (council-place), where anybody could approach him directly with news, petitions, and complaints. He was also the representative and spokesman of the tribe in all its external relations.

The chief was always treated with great respect. He was ceremonially addressed by the personification of the tribal name (e.g., as MoNgwato or MoKgatla), and his deeds were extolled in special praise-poems (*mabôkô*) recited at important assemblies. His installation and marriage were occasions of great public festivity, and his death was universally mourned. He received tribute from all his subjects, in both labour and kind;[11] he had the first choice of land for his home, fields, and cattle-posts; he and his family were leaders of the age-regiments, and took precedence in matters of ritual; and he alone had the right to convene tribal assemblies, create new regiments, arrange tribal ceremonies, and impose the supreme penalties of death and banishment. Offences against him personally were usually punished more severely than if committed against other tribesmen, and disloyalty or revolt against his authority often met with death and the confiscation of the culprit's property. If his own conduct was unsatisfactory, he could be warned or reprimanded by his advisers or at public assemblies; if he ruled despotically or repeatedly neglected his duties, the people would begin to desert him, or a more popular relative would try to oust him by force, or, in the last resort, he might even be assassinated (as happened, for instance, to the Kwena chief Motswasele II in 1821).

Nowadays the chief still performs many of his former duties and enjoys many of his former privileges. But his religious and magical activities have largely ceased since the introduction of Christianity, he no longer leads in war, and his use of tribal labour has been restricted; and although he continues to regulate the distribution of land, his importance in economic life has been greatly diminished by the creation of Tribal Treasuries and Funds. He no longer determines foreign policy, and his criminal jurisdiction has been severely curtailed, especially in the Transvaal. He still retains much civil jurisdiction, and, especially in the Protectorate, is allowed and even encouraged to make local laws for the better conduct of tribal affairs. He is himself subject to the overriding authority of the Government, to which there is an appeal from his judicial and other decisions, and which can fine, suspend, or even depose him if he does not rule satisfactorily. He is held responsible for maintaining law and order in the tribe, preventing crime, and collecting tax and other dues; he must carry out all orders and instructions lawfully issued to him by officers of the Government, both administrative and technical; and he is expected to assist them in all their local activities. His formerly undivided control over virtually every sphere of tribal life has thus been diffused through various Government Departments with superior authority. But, especially in the Protectorate, he remains the central figure of the tribal government, and without his approval and backing no major enterprise among his people can be initiated or successfully accomplished.

The Chief's Advisers and Councils[12]

In running the affairs of his tribe, the chief is assisted firstly by some confidential advisers, whom he consults privately and informally. He chooses and varies them as he wishes, but they usually include his senior paternal uncles and brothers, whose traditional right and duty it is to serve in this capacity. In some tribes he occasionally also consults a wider and more formal body, consisting either of all the ward-heads

---

[10] See below, p. 60.
[11] See above, p. 26.
[12] Schapera, 1938, pp. 72–83 ; 1940 (b), pp. 71–3 ; Ashton, 1937, pp. 76–8 ; 1947, pp. 244–51 ; Lestrade, 1928, pp. 429–32 ; Language, 1943 (a), pp. 26–68 ; Willoughby, 1923, pp. 94–7 ; Breutz, 1941, pp. 95–8 ; Mackenzie, 1871, pp. 371–3.

or at least of those whom he trusts. These men are specially summoned whenever a matter of urgency or importance needs to be discussed; the meeting, which is secret, was formerly held at night in the great cattle-kraal adjoining the tribal council-place, but nowadays it sometimes takes place by day and wherever convenient.

All matters of public policy are dealt with finally at an assembly open to all the men of the tribe, and variously termed *pitsô, lebatla,* or *phuthêgô*. Such assemblies are held very frequently, at times almost weekly, and they usually meet early in the morning in the tribal council-place, close to the chief's residence. Normally only the men present in the capital attend, the business discussed and decisions reached being communicated, if necessary, to the inhabitants of outlying villages through their headmen. But on important occasions the men from outside are also summoned, and if the matter is at all critical (e.g., in case of serious internal disputes) they may even be compelled to attend. A crucial meeting of this kind is sometimes held in the open veld some distance from the capital, and the men all come to it armed and ready for trouble; it is then known as *letsholô*, a name also applied to the collective hunt by which it is usually preceded or followed.

Since anybody present is entitled to speak, the tribal assemblies provide a ready means of ascertaining public opinion. The decisions reached are generally those already formulated privately by the chief and his personal advisers, who because of their standing are able to persuade the others to support them, but it is also not unknown for their wishes to be overruled. The discussions are characterized by considerable freedom of speech, and, if the occasion seems to call for it, the chief or his advisers may even be severely criticized. In the Protectorate it is now statutory law that the chief may not make regulations, impose levies, or use certain forms of tribal labour unless the people have first expressed their approval in *kgotla;* and, both here and in the Union, such assemblies are frequently used by the Government as a means of informing the tribes about new legislation and other developments or of inquiring into local disputes.[13]

In his judicial and administrative capacities the chief again relies mainly upon his personal advisers; and if he is ill, or away from the capital, his heir or some other close kinsman acts as his deputy (an office nowadays officially recognized in the Protectorate). Some of his advisers are specially used to help him receive and judge cases, supervise the execution of his verdicts, and carry important state messages; in the old days they also arranged tribal ceremonies and commanded the army under his direction. These men, variously known as *dintona* (" lieutenants ") or *dikala tsa kgosi* (" the chief's right-hand men "), are seldom closely related to him, and may even be commoners. For minor tasks, such as carrying simple messages, he uses any tribesman at hand. Nowadays he often also has an office staff of clerks, tax collectors, etc., and perhaps several policemen to detect and arrest criminals and to act as messengers of his court. In the Protectorate many tribes further have special committees to look after such matters as finance, education, and agricultural development; the members were at first appointed by the chief, but are now usually elected for specified periods by the tribal assembly.

Local Government[14]

Every family-group, ward, village, and section, in the tribe has a recognized headman, whose position is hereditary. Within his group he has very wide authority. He is responsible for the maintenance of law and order, and adjudicates over disputes between any of his people or involving them as defendants. He organizes and directs their collective activities, and can freely command their services for public

[13] Among the Ngwaketse, minutes of the tribal assemblies have been kept sporadically since 1910. Those for the period 1910–17 have been translated and edited with full commentary by Schapera, 1947 (b).
[14] Schapera, 1938, pp. 89–103 ; 1940 (a), pp. 58–62 ; Ashton, 1937, pp. 70–6, 80–2 ; Breutz, 1941, pp. 94 f. ; Language, 1943 (a), pp. 1–25.

purposes. He must see that they carry out the commands of his political superiors, and formerly also collected the tribute that they paid to the chief. He is their official representative and spokesman, supports and protects them in their dealings with outsiders, and sees that they obtain the land and other rights to which they are entitled as members of the tribe. Formerly he also was responsible for such magical ceremonies as " doctoring " them or their settlement when they moved to a new locality, went to the initiation schools, or suffered from epidemic disease.

In carrying out his duties, the headman is assisted firstly by his near relatives, especially his paternal uncles, brothers, and adult sons; and during his absence or illness, or after his death, the man next to him in rank automatically takes his place. In effect, therefore, the government of any group is vested not so much in one particular person as in the whole lineage of which he is the head; and the senior member of that lineage present on any occasion when action must be taken is able, by virtue of his birthright, to exercise authority over all the other people of the group. The ward-head is assisted also, especially in his judicial and administrative capacities, by the elders of the other family-groups; the village or sectional headman is similarly assisted by the other ward-heads in his village or section, as well as by the men who advise him in his own capacity of ward-head. Finally there is a formal council embracing all the adult men of the group, whom the headman calls together to discuss any matter of importance affecting it as a whole. These advisers, individually or collectively, act as a check upon his behaviour, and can reprimand or even punish him in case of misrule; usually, however, they prefer to complain to a higher authority.

The various forms of local authority are graded. The elder of a family-group is directly subordinate to his ward-head, who in turn is subordinate to his village or sectional headman; the latter, finally, is subordinate to the chief. In each case the superior official has an authority overriding that of the lesser officials in his own group. They must obey his commands and carry out his instructions, they must refer to him all cases which they are unable to settle or with which they are not competent to deal, and there is an appeal to him from all their judicial and executive decisions. The chief, moreover, has the power to punish any headman for neglect of duty, and in case of extreme incompetence or abuse of office may even depose him; the responsibilities of headman then fall to the man next in order of succession. In effect, therefore, the government of the tribe is ultimately concentrated in the hands of the chief, but the existing social and territorial organization is used to delegate matters of more purely local concern to subordinate authorities. General authority over a major group, it may be noted, is always accompanied by special authority over a minor group; a village headman, for instance, is also head of his own ward in the village, and elder of his own family-group within the ward.

In the larger tribes of the Protectorate, there has relatively recently been a new development in the system of local government: for administrative purposes, all the villages outside the capital are nowadays grouped into a few districts, each under a man specially appointed by the chief. The governor's main duties are to communicate the chief's orders and messages to the people under his control, hear appeals from the verdicts of their headmen, settle disputes between neighbouring communities, organize and direct local public works, supervise the collection of tax, and advise the chief on local political and economic conditions. All matters that he cannot himself settle he must refer to the chief, to whom there is also an appeal from his own decisions. Should he prove incompetent, he may be recalled by the chief, someone else being sent to replace him; failing this, his appointment tends to be permanent, and may even become hereditary, unless the chief sees reason to intervene. The men selected for the post are sometimes senior relatives of the chief, but usually also include several trusted commoners.

Law and Judicial Procedure

Tribal Law[15]

Tswana law, in its traditional manifestations, is not formally codified. Many rules of conduct, however, stand apart from all others in that they can be enforced, if necessary, by the powers of physical compulsion vested in the courts. These rules, which may be termed the tribal " laws," consist very largely of usages already established in practice. The role of the courts, in case of dispute, is to confirm their validity, or to decide what the rule really is; the courts thus create a series of binding precedents. Other rules consist of decrees promulgated by the chief after discussion in the tribal assembly. Legislation of this kind did not figure prominently in the old tribal life, but in more recent times, owing especially to the new conditions created by contact with Europeans, it has become a fairly common means of modifying or adding to the existing law. As a result, there now often is considerable variation in the laws of different tribes; usages enforced in some may be specifically prohibited in others (e.g., *bogadi*), and some tribes have laws that are unknown elsewhere. New laws have sometimes been recorded in writing, and several chiefs have also compiled partial lists of the laws passed by themselves and their predecessors;[16] but, in the main, tribal law is still unwritten.

In practice, although not in theory, the Tswana distinguish between " criminal " and " civil " law. The civil law establishes, *inter alia,* the private rights of people in regard to personal status, property, and contracts, and provides for redress, if such rights are violated, by compelling restitution or compensation. The principal civil wrongs recognized in Tswana law include seduction, adultery, and similar offences against family rights; breaches of contract; trespass, damage, theft, and similar offences against property; and defamation and other wrongs against reputation. The criminal law treats certain acts not merely as injuries to specific persons, but as offences harmful to the tribe generally and therefore deserving of punishment. The principal crimes formerly recognized included incest, sorcery, homicide, rape, and other forms of serious bodily assault, offences against tribal authorities acting in their official capacity, and breaches of the laws decreed by the chiefs. Occasionally, as in some types of theft and bodily assault, the offender may be both punished and forced to make amends to his victim; but the initial reaction to his deed is usually sufficient to show if it is regarded mainly as a crime or a civil injury.

Courts and their Jurisdiction[17]

In the old days, as already noted, every local division of the tribe from the ward upwards had its own court, with the headman as judge. Junior courts had full jurisdiction in civil matters, although if the judge found any case too difficult he would himself refer it to a higher court for settlement. Ward and village headmen also heard criminal cases, but usually only in order to ascertain the facts; if the crime was at all serious (e.g., bloodshed or sorcery), they were bound to send the culprit to be tried by the chief, who alone had the power to pass sentence. Appeals could be taken from any junior court to its immediate superior, but the decisions of the chief's court were final. There was, however, a limited right of asylum, except in cases of homicide: a convicted criminal who managed to escape from court to the home of the chief's mother was entitled to pardon.

The old tribal courts still continue to function almost everywhere, but usually only in an arbitral capacity, i.e., they have no legal power to enforce their decisions. The only courts with judicial powers are those officially and specifically recognized by the Government; they consist as a rule of the chief's court and (in the Protectorate) a small number of others each embracing many different wards or even several

---

[15] Schapera, 1938, pp. 35–52, 257–78, etc. ; Krüger, 1935, pp. 112–36 (an attempt to " codify " Tswana legal rules).

[16] Schapera, 1943 (e).

[17] Schapera, 1938, pp. 279–83 ; 1943 (d), pp. 27–40 ; Language, 1943 (a), pp. 124–30, 150–6 ; Ashton, 1949, pp. 715–7 ; Hailey, 1938, pp. 410 f.

villages. The constitution, range of jurisdiction, and methods of procedure of these courts are statutorily defined in the Union under Section 12 of the Native Administration Act, 1927, as amended by Section 23 of Act No. 36 of 1944, and in the Protectorate under the Native Courts Proclamation (No. 33 of 1943); the Protectorate legislation also insists upon written records being kept of all cases heard. As already mentioned, an appeal lies from the chief's decisions to the court of the local Native or District Commissioner; from this court, again, appeal can be made in the Union to the Native Appeal Court (Central Division) and thence, under certain conditions, to the Appellate Division of the Supreme Court, and, in the Protectorate, to the High Court and thence, under certain conditions, to the Privy Council in Great Britain.

In British Bechuanaland, chiefs have civil jurisdiction in all matters except those arising out of Christian or civil marriages, and criminal jurisdiction in all matters except " rape, murder, culpable homicide, pretended witchcraft, and theft from other tribes," but they may not impose any punishment involving " death, mutilation, and grievous bodily harm."[18] In the Transvaal their civil and criminal jurisdiction is limited to specified offences arising out of or punishable under " Native law and custom," and their penal powers are restricted to a fine not exceeding £5 or two head of cattle.[19] In the Protectorate, they may try any case except homicide, civil marriage and inheritance disputes, and (unless specially exempted) sorcery, and they may impose punishments not exceeding two years' imprisonment, eight lashes, or a fine of £100, 25 head of cattle, or 150 small stock; subordinate tribal courts all have lesser powers, which are set out in the warrants under which they operate.[20] Otherwise, in both the Union and the Protectorate, Natives are subject to the ordinary laws of the land (i.e., those made by the Government), and they have the right to be tried by European instead of tribal courts if they wish and can show cause.

Conduct of Trials[21]

Despite the changes noted, cases heard by the tribal courts are still conducted mainly in the traditional manner. Action in regard to civil disputes lies with the person aggrieved. Unless he chooses to ignore the matter altogether, he should first try to reach a settlement out of court; if this does not succeed, he lodges a formal complaint with the offender's ward-head, who will fix a day for hearing the case. A crime cannot be settled out of court, but must always be tried; a preliminary investigation is usually first held by the accused person's ward-head, who will then refer the matter, if necessary, to the court of the chief or some other superior. Courts have no fixed sessions, but meet as required; cases are normally heard as soon as possible after being reported, sufficient time being allowed for the parties and their witnesses to assemble.

All those concerned gather early in the morning of the appointed day at the judge's *kgotla* (council-place). He and his advisers sit facing all the other people, who usually group together indiscriminately, except that the persons involved are in front. Witnesses are not kept away until required, but sit in from the beginning. Any adult tribesman who wishes may attend, all trials being conducted publicly, but women and children are excluded unless parties to the case. Each litigant is supported by his relatives and friends, and is also responsible for producing his own witnesses; if he has duly notified them and they do not appear, the judge sends messengers to fetch them; failure to obey the summons is punished, unless there is a valid excuse, in which case the trial is sometimes postponed if the man's evidence is considered essential.

After the judge has briefly introduced the case, he calls first on the plaintiff and then on the defendant or accused. Both men speak freely and uninterruptedly,

---
[18] Rogers and Linington, 1949, pp. 205, 209.
[19] Ibid. pp. 204 f., 208 f.
[20] B.P. *Annual Report*, 1947, pp. 34 f.
[21] Schapera, 1938, pp. 283–300; Language, 1943 (a), pp. 131–50, 153; Mackenzie, 1871, pp. 373–5.

and after making their statements are questioned closely by anybody who wishes. The witnesses are next heard, starting with the plaintiff's, and they too are questioned in turn. When everybody concerned has spoken, the matter is thrown open for general discussion. All those who wish ask further questions and express their opinions; the judge's advisers, one by one, then debate the issues and merits of the case, referring if necessary (and possible) to precedents; he himself finally summarizes the evidence and the opinions expressed, discusses them, and delivers his verdict. This generally accords with what most of the others have said, but he is free to decide otherwise, although he should then explain why he dissents. If the case is taken on appeal to a higher court, it is there heard all over again from the very beginning. The judge who first tried it must also state how he reached his decision; and occasionally, if he was blatantly wrong or prejudiced, he may be reprimanded or even punished.

The evidence most favoured in reaching a verdict is that of eye-witnesses; circumstantial and hearsay evidence are also admitted, but carry less weight. There is no form of oath, ordeal, or divination. Knowledge of a person's character helps to determine the value placed upon his evidence; the extensive questioning which he undergoes if his word is doubted generally suffices to show if he is telling the truth, and if found deliberately trying to deceive the court he may be punished.

In most civil cases, the plaintiff states what he wants as damages or compensation; if he wins, he may be awarded the whole or merely part of his claim, according to the merits of his case and the will of the judge. No fee is due to the court for hearing the case (or, if necessary, enforcing the payment), nor are there legal costs of any other kind. Crimes are usually punished either by fining or by thrashing. Fines formerly consisted of cattle and small stock; one or more of the animals, when delivered, were slaughtered for the men at the court, the others being kept by the judge. Nowadays the animals are all sold for the benefit of the Tribal Treasury or Fund, which also receives any fines paid in cash. Thrashing, in the form of cuts with a light cane, is administered immediately after sentence is pronounced (unless notice of appeal is given); in the Protectorate, a sentence of more than four cuts must be carried out under Government supervision. It is nowadays also possible, in the Protectorate, for tribal courts to impose sentences of imprisonment, which are served in the Government gaols.

RELIGION AND MAGIC

Religion

Christianity

The official religion of most tribes is now Christianity. Missionaries first reached the Tswana in 1816, and by about 1870 were established in all the larger tribes. They usually succeeded fairly soon in converting the chief, whose acceptance of their teachings both served as an example to others and made him abandon many of the public rituals for which he was responsible. By no means all Tswana profess Christianity,[1] but, almost everywhere, the Christians include the chief and most other leading tribesmen.

It is difficult to generalize about the nature of Tswana Christianity.[2] The denominations chiefly represented are Congregationalist (London Missionary Society), Dutch Reformed, Methodist, Lutheran, Anglican, and Roman Catholic; there are also some Native separatist churches, but they seem to have relatively few adherents. Until fairly recently, many tribes had missionaries of one denomination only, so that the Christianity of, say, the Ngwaketse (Congregationalist) differed in some features from that of the Kgatla (Dutch Reformed) or Malete (Lutheran). Now that denominational monopolies have largely disappeared, the Christians of even a single tribe often differ in ritual and belief. It may be assumed that, in general, the members of a particular Church profess the doctrines that it preaches. But observations in the Protectorate suggest that relatively few really understand the principles of Christianity, or sincerely try to lead a devout Christian life. Their conceptions of God and Jesus, of salvation and resurrection, and of such rituals as communion, are often very distorted versions of what they are taught, their attendance at Church is largely a matter of routine, they practise in secret many of the old customs forbidden to them, and in various other ways show that their Christianity is merely conventional. The missionary himself is now perhaps better described as a parish priest than as an apostle. New members of the Church are recruited mainly from the children of Christian parents, and their acceptance of Christianity seems to be largely a matter of course instead of genuine conversion.

Christianity has brought much more to the Tswana than merely a new set of religious beliefs. The missionaries built churches, introduced the vocations of preacher and catechist, established local Church councils, instituted new ceremonies (e.g., baptism, confirmation, and communion) and the observance of the Sabbath and other religious holidays, developed new forms of marriage and death ritual, and through their hymns provided a new and very popular form of music. They sought to impose a new system of morality conforming to Christian ideals, and to this end introduced sanctions of various kinds governing the lives of their members. They also taught the people to wear European clothing. For many years they alone were responsible for education in literacy, building schools and importing teachers; more recently, in some tribes, they have carried on medical work and built hospitals, fostered youth movements and other social and recreational activities, and published literature in the vernacular. Some of these developments have created new social distinctions, e.g., between Church members and " heathens "; others have affected the population generally, and it is safe to say that virtually all Tswana are now familiar with at least some manifestations of Mission activity. The missionary himself has become not only the tribal priest, but also the guide and adviser of the people in many spheres of life remote from religion.

[1] See above, p. 47.
[2] Schapera, 1934 (c), pp. 52 f.; 1940 (a), pp. 33–5, 170 f., 264 f., 306–10.

Traditional Cults

In pre-Christian times, so far as can be gathered, the Tswana believed in a high god named *Modimo*, who was regarded as the creator of all things and the moulder of destiny. He was vaguely associated with the phenomena of the weather, and punished innovations or departure from established usage by sending wind, hail, or heat, and withholding the rain; and death, if not attributable to sorcery, was spoken of as an " act of God." He was, however, considered too remote from the world of man to be directly approached in prayer, although at times the ancestral spirits might be implored to intercede with him.[3] There were also many myths about *Dingwe,* a cannibalistic ogre against whom children were specially protected by charms, and about *Lôôwê, Tintibane, Matsieng,* and *Thobêga,* "demi-gods" associated with caves and engraved footprints on rocky outcrops (as in the Kwena and Kgatla Reserves, B.P.). Offerings of meat, corn, and beer were occasionally made to them, together with prayers for rain, fertility of crops, and success in war, but on the whole they seem to have been of minor importance.[4]

The dominant cult was the worship of the dead.[5] Men dying at home were buried in the cattle-kraal of their family-group or ward, and women in the backyard of their compound. The corpse was laid out by some elderly people under the direction of the deceased's maternal uncle, whose duty it was "to handle the putrefaction of his sister's children." A man was buried with his weapons in his hands, a woman with her hoe and some seeds of every cultivated grain; both were also dressed in the clothes they wore when alive. "They were then equipped for their journey to the world of the dead, where a man would continue to herd his grandfather's cattle and a woman to cultivate the soil." Finally, the corpse was wrapped in old skins; for a man of some importance, the wet skin of an ox specially killed for the purpose was used instead. The grave was a round hole, with a niche to one side, in which the body was placed in a crouching position, with the head facing west, " so that he was ready to get up and walk to the place where the souls of the other dead had gone."

As these usages indicate, the Tswana believed in the survival of the dead. They held that the souls of dead people became spirits (*badimo*), which ultimately found their way to a world vaguely located somewhere underground. Here they led a life very similar to that on earth. But they continued to take an active interest in the fortunes of their living descendants, over whose behaviour they exercised a powerful control. They rewarded with good health and prosperity those who treated them with becoming respect and obedience, but punished with sickness, economic loss, or some other misfortune those who neglected them or who offended against the prevailing social code, of which they were the guardians. Hence, in order to retain their favour, they had to be specially propitiated.

Each family was held to be under the direct guidance of its own agnatic ancestors, who in turn were interested only in the affairs of their own descendants. They retained the individual characteristics they had possessed while alive, and their importance as deities was determined mainly by the status they had enjoyed on earth. In practice worship was generally directed only to those more recently dead, such as a father or grandfather, whose personal peculiarities were known and remembered. The head of the family, as senior living representative of the ancestors, conducted the rites. He sacrificed and prayed to the dead at their graves whenever they revealed themselves through dreams or calamity, or in some other form that the diviners interpreted as a sign that they were offended. He also invoked them on all important domestic occasions, e.g., birth, marriage, or the undertaking of some new enterprise, when he offered them libations of beer or sacrifices of fowls, goats, and in emergencies even cattle, and prayed to them for continued guidance

---

[3] Mackenzie, 1871, pp. 394 f. ; Brown, 1926, pp. 113–25 ; Language, 1941, pp. 271–8 ; Smith, 1950, pp. 116–22.
[4] Brown, 1926, pp. 101–12 ; Willoughby, 1932, pp. 36–40, 74–7.
[5] Brown, 1926, pp. 97–108 ; Willoughby, 1928, pp. 40, 57, 99–102, 275–6, 365–6, etc. ; Language, 1941, pp. 278–84.

and help, or thanked them for the blessings they had sent. His role of family priest gave him considerable authority over his dependants; they could not approach their ancestors except through his agency, so that if they quarrelled with him they were cut off from the deities controlling their welfare. A special ceremony of reconciliation was then necessary before he would again sacrifice on their behalf.

Apart from worshipping their own ancestors, all members of a tribe acknowledged those of the chief as a source of welfare and prosperity: just as he and his relatives guided their fortunes on earth, so his ancestors were held to afford supernatural protection and assistance to the people they had once ruled. Hence, on all important or critical occasions, he would sacrifice and pray to them on behalf of the tribe as a whole. The role he thus played as tribal priest—a role which only he, as ruling chief, could fill—helps to explain the great reverence with which he was regarded by his people. He himself became a tribal god after his death. He was buried in the great cattle-kraal adjoining his council-place, and, especially if he had been renowned or well beloved, his descendants would go there with their people to pray to him for rain and harvest, peace and prosperity.

The ceremonies which the chief conducted or sponsored on behalf of the tribe included the doctoring of the army (*go fôka marumô*) after his father's death or in time of war, the consecration of the capital (*go thaya motse*) whenever it was moved to another site, the doctoring of the tribal boundaries (*go bapola lefatshe*), especially in times of pestilence, drought, or threatened invasion, the initiation of boys and girls into membership of age-regiments (*bogwêra* and *bojale*), the sowing, first-fruits, and harvest, festivals (*letsema, molomo,* and *dikgafêla*), and, above all, the making of rain (*go fetlha pula*).[6] Rainmaking was everywhere held to be an attribute of the chiefship, and a chief's reputation and popularity were often determined by the nature of the rainfall during his period of rule. Some of the rites he performed himself, others were carried out at his request or under his supervision by professional rainmakers (*barôka ba pula*) and other members of the tribe.

The chief had a special rain enclosure (*segotlwana sa pula*) behind the hut of his great wife, where several pots of rainmaking medicines were kept. Every year, before the cultivating season started, the immature girls ceremonially brought water to fill the pots; they and the immature boys were then sent to sprinkle some of the contents over the fields and cross-roads near the capital. If the rain did not come, driftwood and other objects connected with water were burned in the enclosure, so that the smoke should summon the clouds; sometimes, too, men were sent to capture alive a wild animal of a specified species and sex, which was then slaughtered, portions of its flesh being added to the mixture in the pots. Failing this, the women would gather at the graves of the chief's ancestors and, sprinkling them with water and beer, would sing special songs of prayer for rain. Ultimately the chief himself would go to one of the graves, accompanied by the people; there an unblemished black bull was slaughtered, and portions of its meat, the skin, bones, etc., were burned on the grave; and as the smoke rose, he would pray to his ancestor " to let the rain fall." As a last resort, search would be made for objects out of place (e.g., a pot hidden in a tree), which were thought to have been deposited by sorcerers to keep off the rain; such objects were doctored and thrown into a river bed or pool, and the people then gathered at the tribal council-place to be sprinkled with the contents of the rain pots and freed from contamination. Throughout the rainy season, too, all newly bereaved people were reported to the chief, on whose instructions they were smeared with the juice of irritant bulbs used as rainmaking medicines; this treatment was intended to " cool " their bodies and prevent them from scorching the land wherever they went.[7]

[6] Willoughby, 1905, pp. 302–5, 308–11; 1928, pp. 226–34, 251–4, 291; Mackenzie, 1871, pp. 383 f.; Language, 1941, pp. 284–9, 307–9, 322–7; Scroggie, 1946, pp. 82–8; Schapera, unpublished notes.

[7] Mackenzie, 1871, pp. 384–8; Brown, 1926, pp. 129–32; Willoughby, 1928, pp. 203–12; Language, 1941, pp. 309–18; Schapera, 1930 (b), pp. 211–6, and unpublished notes.

Modern Survivals[8]

The old religious beliefs and practices are nowadays seldom encountered. Most of the people who do not profess Christianity seem to lack any formal system of worship. Such vague beliefs as they hold are largely derived from echoes of missionary teaching blended confusedly with traditions of ancestor worship, and the Christian conception of God has almost completely displaced the old conception of *Modimo* (whose name the early missionaries adopted for the Biblical deity). Magicians occasionally still pray to the spirits of their ancestors, especially when preparing medicines or performing an important ceremony, but even they have abandoned animal sacrifice and many of the other rituals that were formerly part of the cult. Otherwise it is usually only in sickness that the influence of the ancestors continues to receive recognition. Diviners sometimes attribute a person's illness to the anger of his ancestors, whom he has offended by quarrelling with a senior relative; that relative must then be called to wash the patient with certain medicines and pray for his recovery.[9] Even Christians often practise the rite, although well aware that the prayers are addressed to the ancestors.

Beliefs concerning the fate of the dead vary greatly. Influential men are occasionally still buried in their cattle-kraals, but all other people, except very young children, are now buried in special graveyards on the outskirts of the villages, and whenever possible in wooden coffins. A religious service is held at the graveside for Christians, but there is no corresponding rite for the heathens. Professing Christians have accepted the Church doctrine of immortality, with the associated conceptions of Heaven and Hell. Some of the heathens retain the ancient belief that the soul goes underground to a world of the dead, but they seldom claim to know what happens to it there; others maintain that it is simply blown away by the wind, so that when a person dies he is annihilated, just like an animal. Both heathens and Christians also believe in *dipoko* ("spooks" or ghosts), a conception almost certainly borrowed, as is the name itself, from Afrikaans folklore. They say that if a person, after death, regrets the cattle and other wealth that he left behind, he will haunt the places that he frequented while alive. His soul rests in the grave by day, but emerges at night and wanders about, in human or animal guise, seeking its lost treasures. People are afraid of these ghosts, and will therefore usually avoid a graveyard at night; there seem to be no other customary usages regarding them.

Almost all the great ceremonies formerly conducted by the chief have been widely abandoned. But, as already mentioned, a few tribes still retain the old forms of initiation, and in some rainmaking is still practised secretly in times of drought. During the early part of the agricultural season, too, certain traditional taboos are commonly observed to prevent the rain from falling as hail. In many tribes the missionaries have introduced an annual day of prayer for rain, and a Church day of thanksgiving in place of the old harvest festival; these ceremonies are attended by Christians and heathens alike, and constitute the nearest approach there now is to a common system of worship.

MAGICIANS AND MAGIC

Magicians

In contrast with the old religious system, magic still persists strongly, and is often employed even by professing Christians. Tswana magic consists essentially in the use of "medicines" (*ditlhare*) for the attainment of certain specific ends which cannot be achieved by empirical methods alone.[10] There are medicines for treating disease, protecting people and their possessions, promoting the fertility of women, livestock, and fields, ensuring success in courtship and hunting, finding

---

[8] Schapera, 1936, pp. 245, 248 ; 1940 (a), pp. 306–10
[9] Willoughby, 1928, pp. 176, 194 ; Schapera, 1934 (a), pp. 298–305.
[10] Mackenzie, 1871, pp. 381–3 ; Brown, 1926, pp. 126–32, 136–40 ; Schapera, 1938, pp. 255 f., and unpublished notes.

work, injuring enemies, and many other purposes. Their use in such contexts is normally termed *go alafa*, " to doctor." Some types of magical activity are more specifically distinguished, e.g., *go thaya*, to " fortify " people and places against hostile influences, *go upa*, to expel or keep away garden pests, *go fetlha pula*, to make rain, and *go lôwa*, to bewitch.

Certain medicines and their uses are known almost universally, e.g., the remedies for minor ailments, the smearing of a threshing-floor with *sekaname* to prevent loss of grain, the scattering of *mogaga* peels by newly bereaved people to " cool " the ground on which they tread, and, in a more limited context, the love medicines favoured by boys. Whenever the need arises, the person concerned acts as his own magician; he seeks or makes the appropriate medicine for himself, and uses it in the recognized manner. In the rare event of his not knowing what to do, he will readily and freely be told by someone else.

Most forms of magic, however, are known only to specialists, part of whose livelihood is gained from the hire of their services. These specialists, collectively termed *dingaka*, " doctors " (sing. *ngaka*), are classified broadly into *dingaka tse dinaka*, " horned doctors," and *dingaka tse ditshopya*, " hornless doctors." Both have as their main work the treatment of disease, but whereas the former diagnose by means of divining-bones (*ditaola*), the latter do not practise bone-throwing and rely upon personal examination of the patient. Most doctors are also proficient in other kinds of magic, but very few claim to be versed in every branch, and some of the techniques (e.g., for treating women's menstrual complaints, removing garden pests, avenging death due to sorcery, and making rain) are highly specialized.

Both men and women may become doctors, but in fact very few women adopt the profession. It is often associated with certain families, within which it is handed down from parent to child. But anybody else who wishes may also learn, by apprenticeship to some doctor of his acquaintance, to whom he pays a fee of one or more head of cattle. In some tribes entrance into the profession was formerly controlled by the chief; whoever proposed to become a doctor required his permission, and sometimes he himself, with the aid of his tribal doctors, selected promising youths and saw that they were duly trained. The initial period of instruction usually lasts from one to three years, during which the apprentice learns all that his teacher knows or wishes to impart; he is then doctored to give him the power of working on his own. Thereafter he may start practising at once, but usually he first goes to other doctors to widen his knowledge and experience.

The fees charged by doctors vary according to the nature of their services. The standard rates are 1s. or 2s. in cash, a goat-skin, or some corn, for divination, a sheep or a goat for minor cases of illness, and a heifer or some cash equivalent for serious ailments or charming huts, cattle, etc. Divinations are paid for in advance, " to make the bones speak "; the other fees are not due unless the treatment has been successful, and if payment is withheld the doctor may have to sue for it in court.

In the old days, doctors renowned for their skill were generally employed to help the chief in the various public rites for which he was responsible. Such men were called *dingaka tsa morafe*, " tribal doctors." Rainmakers always belonged to this class, as did the doctors who treated the army in time of war, assisted at the initiation ceremonies, purified newly bereaved persons, etc. They carried on private practice in the same way as others, but the chief had first call on their services, for which they were usually well rewarded. With the abolition of most tribal rites, they have lost much of their former importance. The chief still has his own doctors for various private and public purposes. Those distinguished in this way are commonly recognized as the leaders of their profession, but have no special power in regard to their fellows; there is no organized body of doctors, they never meet collectively for ceremonial or other professional purposes, and they are subject only to the ordinary political authorities. Doctors convicted of malpractice (e.g., sorcery or inducing abortions) may be punished and ordered to suspend or abandon their professional activities, and if aliens may be banished from the tribe. In the Protectorate, the " Witchcraft Proclamation " (No. 7 of 1927) makes it a penal offence for doctors to practise divination, or for people to employ them to identify sorcerers, but these regulations are seldom enforced.

Morphology of Magic[11]

Tswana magic consists basically of (*a*) medicines, and (*b*) their use in a particular manner; sometimes a spell or prayer is added. The medicines, as suggested by the generic name *ditlhare* (sing. *setlhare*, tree), are mainly of vegetable origin, i.e., roots, bulbs, bark, wood, or leaves, of trees, shrubs, and plants; portions of the flesh, skin, bones, or blood, of animals of all kinds (including human beings) are also commonly used, and, occasionally, mineral powders (e.g., hæmatite). In the simpler forms of magic, known to almost everybody, only one substance may be used, which does not require special preparation; a raw *sekaname* bulb, for instance, suffices to protect a threshing-floor. But, for most purposes, the medicine consists of many different ingredients which have to be carefully blended. Thus, *mothusô* (for fertilizing crops) is compounded from the roots of six different trees and the twigs of another, fragments of hippopotamus and ostrich bone and of ostrich and python flesh, feathers of the flamingo and sea eagle, berries of a certain species, and two varieties of sunflower seed, all of which are burned together in a potsherd, ground to powder, and mixed with fat; and the ingredients of *leswalô* (burned to obtain good fortune) are all cut to pieces, dried in the sun, pounded fine, mixed with fat, and worked into balls, from which fragments are detached as required.

The use of the correct medicines is considered essential for the success of any magical activity, and it is to their specialized knowledge of medicines that doctors owe their powers and livelihood. The efficacy of a medicine derives from the fact that it is traditionally employed for a particular purpose; the first doctors, it is said, were taught by God (*Modimo*), and their knowledge has been handed down from one generation to another. Hence it is only by instruction, and sometimes in dreams (where they commune with their ancestors), that people can learn about medicines and their use. Doctors wishing to enlarge their repertory apply to colleagues with a more comprehensive training; it is inconceivable that they should resort for the purpose to experiment. Many medicines or their ingredients can be obtained locally, the doctor gathering them in the veld when new stocks are needed; occasionally he plants specimens of the rarer growths in his compound, so that supplies are always available. Some medicines, however, occur only in distant regions (e.g., sea-weed and other maritime products used in making rain); these the doctor obtains by making special trips abroad, or by purchase from his colleagues, especially visitors coming from regions where they abound.

Magic depends for its efficacy not only upon the use of the right medicines, but also upon their use in the right manner; *sekaname*, for instance, must be smeared on a threshing-floor in the form of a St. Andrew's cross. In some of the more complex rituals, a general pattern must be followed, but the details are not stereotyped. Thus, in doctoring a new cattle-kraal, the medicine may be smeared onto pegs, or on a stone, and buried in the ground, or it may be smeared on the poles of the fence; the rite varies with the doctor, but so long as the medicine is left somewhere in the kraal it will be effective. In the treatment of disease, and in other treatments extending over some time, the doctor may perform the initial rite only, afterwards leaving medicines behind with instructions for their future use; sometimes he himself does nothing at all, but simply sells the medicine to a client and tells him what to do. The layman using medicines rarely has to observe any special conditions except those prescribed as part of the rite; thus, nobody may work in a field on the day when it is doctored for the removal of pests. But doctors, when preparing their medicines or about to perform a major rite, must refrain from sexual

---

[11] Schapera, 1934 (b), pp. 564-77, 582-4, and unpublished notes ; Scroggie, 1946, pp. 73-8, 128-31.

relations and observe certain other taboos to prevent the medicine from becoming " spoiled "; they do no magic at all on the day after someone has died in their neighbourhood; and, if the dead person was their own patient, they also carefully purify all their medicines before resuming work.

In the simpler forms of magic, there is no accompanying verbal formula; the correct use of the correct medicines is normally considered sufficient. In the more complex rites, an audible spell or prayer is often added towards the end. The spell (*tiisô*, from *go tiisa*, " to make firm ") gives additional potency to the medicines. Its words indicate the purpose of the rite, and may also instruct the medicines to be efficient; but the formula is not rigid and can be varied on different occasions, nor is it the treasured secret of its user. Many doctors prefer instead to pray directly to the spirits of their ancestors, whom they ask to bless the medicine and make it work; they maintain that the ancestors are its " owners," and therefore the best guarantee of its success.

Divination

Most doctors, in addition to their other activities, practise divination (*go laola*). This features prominently in Tswana life; people use it to discover the nature and causes of a sickness, the reasons for a person's death, the whereabouts of missing stock, the prospects of a journey, the meaning of unexpected objects seen about the compound, and in all other situations where they are baffled by some occurrence or wish to ascertain what the future holds in store. In the old days, similarly, the chief consulted diviners officially before holding any big tribal ceremony, in time of war and drought, when selecting the site of a new village, etc.

Various methods of divination are known, but by far the most common is the use of " bones."[12] These consist of two sets, generally used in combination—(*a*) The *tlhabana* are small, thin, rectangular tablets, decorated on one surface and plain on the other; there are four of them, two representing males and two females (old and young of each sex) and (*b*) The *bola* are astragalus bones of animals (ideally there should be two from every common species of quadruped, one male and one female), together with such other objects as tortoise scales and shells of certain species. The two sets are kept together in a small skin bag; when required for use, they are gathered in both hands, told the purpose for which they are being consulted, and thrown to the ground. The throwing is always done by the consultant himself; the doctor scrutinizes the position in which the bones have fallen, recites the " praise " conventionally associated with it, and then proceeds to interpret it.

The system of interpretation is based upon the primary significance of each " bone," the surface on which it falls, the direction in which it points, and the relative disposition of all the " bones." The *tlhabana* tablets, for instance, can each fall with its decorated surface either exposed or concealed, so that, collectively, they can assume any one of 16 different combinations. Each combination has its own name, its own " praise," and a stereotyped general significance; thus, if the tablets all show their decorated surfaces, the combination indicates turmoil, confusion, and assembly, whereas if they are all reversed the indications are burning, destruction, and loss. The astragalus bones and other *bola* serve mainly to fill in detail, and allow for greater flexibility of interpretation within the general pattern. Tswana who have attended many divinations are usually familiar with the broad principles of " reading the bones," and the role of the doctor often consists mainly in adapting the stereotyped significance to the needs of a particular situation.

Tswana also attach importance to omens;[13] they believe that certain events (e.g., a ringing in the ears), or the sight of certain objects (e.g., a shooting-star), have special significance. Normally omens such as these are not taken seriously; but if an animal behaves out of character (e.g., if a hen crows like a cock), it is

[12] Brown, 1926, pp. 151–4 ; Willoughby, 1928, pp. 136 f. ; Schapera, unpublished notes.
[13] Mackenzie, 1871, p. 392 ; Brown, 1926, pp. 92 f. ; Schapera, unpublished notes.

usually killed at once, lest the owner fall sick and die. Dreams also are regarded as ominous, and people may consult others to ascertain their meaning if not already known; but, even if the omen is unfavourable (as in dreams of raw meat, water, or the chief), apparently no attempt is made to avert it.

Sorcery[14]

Magic is often used maliciously to injure people or their property. It is then termed *boloi*, " sorcery," the practitioners themselves being *baloi* (sing. *moloi*). Tswana distinguish between *baloi ba motshegare*, " sorcerers by day," and *baloi ba bosigo*, " sorcerers by night." Both practise sorcery deliberately and consciously, from motives of greed, vengeance, or envy; there is no belief, like that so widely held in other parts of Africa, that their activities are due to some innate physiological condition beyond their control. The " sorcerers by day," who do actually exist, use magic solely in order to kill or injure a specific victim, and once they have achieved their purpose are usually content. The " sorcerers by night " seem to be largely fictional. They are described as consisting mainly of elderly women who make a habit of bewitching, and many weird beliefs (not all taken seriously) are held about them and their activities; it is said, for instance, that they gather at night in small groups, either naked or wearing very little, and visit all the compounds whose inmates they wish to harm; that they have special medicines for unlocking doors, making people sleep heavily, and exhuming newly buried corpses; and that the owl and hyena serve respectively as their sentries and steeds.

All sorcerers use medicines, generally purchased from doctors. Doctors themselves may be hired as agents, or may try to harm their own enemies by magic; they, too, are then regarded as sorcerers. Various techniques are employed, e.g., concealing some " doctored " substance in the prospective victim's home, sprinkling doctored blood in his compound, treating dust taken from his footprint, or inducing a wild animal to maul him. The one most commonly mentioned, and most feared, is *go jesa*, " to feed." This consists in putting medicines into the victim's food. The medicine may be some such substance as crocodile brains (which when swallowed " changes into a miniature crocodile that gnaws away at the victim's entrails until he dies in agony "); sometimes, however, genuine poisons are used.

There are several methods of protection against sorcery. Thus, almost all people have their bodies magically " strengthened " by a doctor, who inoculates them with the medicine known as *tshitlhô;* this not only renders them immune, but will react disastrously upon whoever tries to bewitch them. Some take the further precaution of swallowing prophylactic medicines before eating away from home; then, if they are given poisoned food, the medicines will make them vomit in time. New huts and cattle-kraals are also " fortified " by means of doctored pegs, stones, etc., which are buried in the ground or placed in some other way; should a prospective sorcerer then enter, the medicine he is carrying will immediately become ineffective, or may even turn against himself.

In the old days, when a person had died, fallen ill, or been afflicted with some other misfortune, a doctor was always called to divine the cause. The bones occasionally attributed the event directly to the action of God (*Modimo*), in which case nothing could be done. Alternatively, and far more commonly, the misfortune might be ascribed to the ancestral spirits, who had then to be appeased by prayer and sacrifice. Often enough, however, the " bones " would indicate sorcery. The doctor, in such cases, never named a person as responsible, but would specify his sex, totem, skin complexion, and the direction from which he usually came when bewitching. If the people could from the description identify the sorcerer, they tried to obtain further evidence of his guilt. Then, if satisfied, they would report him to the chief, who alone had the right to prosecute and to punish sorcerers.

[14] Mackenzie, 1871, pp. 388–91 ; Brown, 1926, pp. 132–6 ; Schapera, 1934 (a), pp. 293–6 ; 1938, pp. 275–8 ; 1952 (a), pp. 41–50 ; Scroggie, 1946. pp. 78–81.

The trial of an alleged sorcerer was usually conducted in the same way as any other case. If he was convicted, he would be ordered to "undo" (*go dirolola*) his victim, i.e., make him recover; should he succeed, he might be let off with a severe reprimand, although sometimes he was also removed from his ward and placed somewhere else. But if the victim died, or was already dead when the case came to court, the sorcerer would be killed; he was taken to the edge of a steep precipice, clubbed or speared, and tumbled over to lie at the base as food for the vultures and hyenas. Nowadays chiefs can no longer impose the death penalty, nor, except in certain tribes of the Protectorate (Ngwato, Ngwaketse, and Kgatla), have they even the right to try a person accused of sorcery. Nevertheless, such trials are still often held, especially if the victim is still living. In them, the bare findings of a diviner are never considered sufficient evidence of guilt; the court must also be satisfied that the accused has in fact resorted to one or other of the recognized methods of bewitching. Great importance is attached to any medicines that may be found in his possession, and unless he can show that they are innocuous he is sure to be convicted. On the other hand, should the accusation appear malicious and unfounded, the accused may be awarded damages for the "spoiling of his name."

The sentence imposed upon a convicted sorcerer varies according to circumstances. Occasionally, if he is found with medicines which he cannot explain away, he is made to swallow them, the argument being that if they are not poisonous no harm will befall him, whereas if they are he deserves the fate that he will suffer. Alternatively, he may be fined or thrashed, or, as in the old days, he may be ordered to "undo" the victim, the implication being that drastic action will be taken against him should the latter die. If the victim is already dead, and there is no direct evidence of poisoning that could be submitted to the Government authorities, his relatives may be given permission to resort to vengeance magic, by doctoring his grave so that the sorcerer also dies. But the most common punishment is to remove the sorcerer from his home. This arises from the fact that he and his victim are usually very closely related and living together, e.g., as husband and wife, parent and child, or brother and brother; less commonly, they may be parent-in-law and child's spouse, master and servant, or doctor and client, but it is extremely rare for people to be accused of bewitching either strangers or persons living far away. It is held, therefore, that the best way of dealing with a sorcerer is to send him where he will be unable to attack his victim again.

TSWANA TRANSFORMATIONS, 1953–1975

Supplementary chapter by John L. Comaroff

Since this volume was originally published, the politico-administrative environments of the Tswana have undergone major changes. The period has been marked, particularly, by the emergence both of independent Botswana and of Bophuthatswana, the "ethnic homeland" created by the South African regime as a consequence of the policy of *apartheid*. These processes have, in turn, been accompanied by efforts on the part of the two governments to initiate social, economic and political transformations among the rural communities—albeit with different ideological justifications. Although anthropologists have documented aspects of modern Tswana life, most recent research has concerned specific problems rather than the analysis of general trends; as a result, the ethnographic picture is extremely diverse. Thus, for example, while one study[1] notes a remarkable continuity in the structure of the Kgatla ward, another[2] describes a dramatic proliferation of new social forms among the Tlhaping. Similarly, in examining the relationship between traditional Tswana systems and contemporary politics in South Africa, writers have stressed widely contrasting perspectives.[3] In short, the situation is apparently one of great variability and complexity. In order to identify and account for overall trends, I shall begin by outlining the development of national policy, with special reference to rural government, during the period under review; for this provides a convenient starting-point from which to comprehend local transformations.[4]

THE NATIONAL CONTEXT

(i) *South Africa*

The accession to power of the Nationalist Party in 1948 was followed by the formal introduction of *apartheid* or "separate development." While its territorial basis had been laid by the *Natives Land Act* of 1913,[5] the elaboration of this ideology into a system of government began with the passing of the *Bantu Authorities Act* of 1951. This Act made provision for the creation, in rural black South Africa, of an administrative hierarchy with three levels: (*a*) each community would be governed by a *tribal authority*, usually headed by the local chief and composed largely of his councillors; (*b*) representatives of the authorities of one district would combine to form a *regional authority*, which was to meet with the Bantu Commissioner, generally under the chairmanship of the head of the largest constituent tribal unit; and (*c*) the "homeland" of each ethnic grouping would be administered by a *territorial authority*, the composition and functioning of which was still to be specified. The Natives' Representative Council (p. 49 above), which had not met since 1948, was simultaneously abolished.

Ostensibly, this legislation would establish a modern order founded upon indigenous government institutions.[6] Its premises have, of course, been the subject of

---

[1] Schapera and Roberts, 1975: *passim*.
[2] Pauw, 1960(b): *passim*.
[3] Cf. Breutz, 1958, Jeppe, 1974 (in Vosloo, Kotzé & Jeppe, see fn. 7) and E. S. Moloto, "The significance of the tribal system in the development of the Homelands." *Race Relations News*, **36**, 3 & 4, Mar. & Apr. 1974, with J. L. Comaroff, 1974.
[4] In order to retain the scope of the original volume, discussion here is limited to the rural chiefdoms. It is to be noted, however, that there is a conspicuous lack of studies of urban Tswana communities.
[5] For a discussion of this Act and its implications for the indigenous peoples, see Sol T. Plaatje, *Native Life in South Africa*, n.d.
[6] A more detailed analysis of rhetoric and reality in this respect is provided in J. L. Comaroff, *loc. cit.*

intense political debate,[7] not least amongst the rural Africans who were requested to discuss it. In theory, they could accept or reject the Act. Tribal records, however, indicate that, while many Tswana groupings opposed it, most concurred for explicitly pragmatic reasons.[8]

Between 1951 and 1959, little was done to implement the *Bantu Authorities Act* : The government appears to have been engaged primarily in the further formulation of policy. Thus a *Commission for the Socio-Economic Development of the Bantu Areas* (the Tomlinson Commission), established to consider the problems of planning within the rubric of *apartheid*, presented its report in 1954. Although the government rejected some of its recommendations (e.g., the introduction of individual land tenure in order to create " a true Bantu farming class "), this report provided a blueprint for the construction of the homeland system.[9]

The *Promotion of Bantu Self-Government Act* (1959) marked the next stage in the emergence of the system. Apart from terminating all parliamentary representation for blacks (p. 49 above), this Act recognized eight ethnic or " national " units; thus the Tswana reserves now became their official homeland. Broadly the same territory as that scheduled in the *Natives Land Act* of 1913, this " bantustan " consists of six large and some twenty smaller disconnected blocks of land with its capital in Mafeking. It has a population of 610,529,[10] 80% of whom live in some 72 chiefdoms.

The Tswana Territorial Authority (T.T.A.), or homeland government, was constituted in 1961 (Proclamation No. 585). Until it was granted executive powers (Proclamation No. R.141 of 1968), the T.T.A. met as an advisory body concerned with practical arrangements for Bophuthatswana, as the homeland was to be called. Thereafter, the T.T.A. could pass enactments, but not laws. In 1971, however, the *Bantu Homelands Constitution Act* replaced the T.T.A. with a Legislative Assembly (T.L.A.), which could legislate with presidential assent.[11] This body has 24 elected members;[12] the remaining 48 being designated by the regional authorities. The T.L.A. elects its own Chief Minister, who must be a chief, and he selects an executive council (cabinet) of five, not more than three of whom may be chiefs. Each heads a department which is responsible for one or more administrative portfolios. Control of the cabinet is at present held by the Bophuthatswana National Party. An opposition, the Seopesengwe Party, was formed recently but has not yet gained any major electoral success.

In theory, the homelands are on the road to independence. But the geo-political implications of this are not entirely clear. For the South African government retains ultimate control over such spheres as external relations, the constitution, currency and security.

Although some 20% of the population of Bophuthatswana live in urban areas situated within the homeland, no standardized system of urban local government has yet been introduced for them. The black residents of the " white " towns, however, are articulated into the overall scheme. The government has ensured that they will

[7] For a justification, see W. B. Vosloo, D. A. Kotzé and W. J. O. Jeppe, *Local Government in Southern Africa*, 1974, Chs. I & II. Popular black attitudes are reflected in M. B. Yengwa, "The Bantustans"—South Africa's "Bantu Homelands" Policy, in A. La Guma (ed.), *Apartheid*, 1972.

[8] See, for example, *Tribal Meeting Minutes*, Barolong boo Ratshidi tribal office, Mafikeng. Also, M. B. Yengwa (*ibid.*: 89f). Some writers (e.g. M. Horrell, *Legislation and Race Relations*, 1966) suggest that, while urban blacks rejected the Act, some chiefs and their supporters might have welcomed it. For a different view, see J. L. Comaroff, *loc. cit*.

[9] The response of the government to the recommendations of the Commission are contained in a White Paper published in April, 1956.

[10] Population Census, 1970. The total number of Tswana in South Africa is given as 1,719,367.

[11] When South Africa became a republic in 1961, the President took over from the Governor-General as Supreme Chief of the African population.

[12] There are twelve regional authorities; a region is also an electoral constituency.

not become a permanent urban proletariat by legislating that *every* black must register as the citizen of a homeland.[13] While the high rate of labour migration to the towns has had an increasingly disturbing effect upon rural social patterns, a possible influx of urban Tswana in the opposite direction is held indigenously to constitute an equal threat for the future. But it is still too early to evaluate whether such fears have any grounds.

The period 1953–75, then, saw the emergence of a new system of local government which gives rural expression to the policy of *apartheid*. Its protagonists claim that it provides a framework for independent black development and harmonious race relations; its opponents, that it is simply a means for extending and propagating white dominance more effectively. But all would agree that it has initiated important transformations in modern Tswana life.

(ii) *Botswana*

It is primarily in the latter years of the period under review that the most significant developments occurred in Botswana : the earlier phase, 1953–1965, culminated in the winding down of the colonial administration and preparations for independence. During this phase, there was little manifest change in either the system of local government or national policy (see above, p. 49f.). The *African Administration Proclamation* (1956) formally recognized tribal councils as chiefly advisory bodies and established, within each chiefdom, subordinate councils under local headmen. But this merely gave official sanction to pre-existing arrangements.

The emergent post-colonial political system, however, involved a fundamental redistribution of authority. This system was founded upon the Westminster model : a multi-party democracy, it has a 35-seat legislature, 31 of which are elected by a popular franchise within constituencies; the government is controlled by the majority party, with executive powers vested in the president-in-cabinet. Outside of the advisory House of Chiefs, the participation of traditional authorities in national politics is specifically barred. Thus a chief or headman who wishes to become an active politician must first relinquish his office.[14]

It was not only at the national level that the role of traditional authorities was redelimited. By the time Botswana became an independent republic, a Local Government Committee (1963) had already recommended substantial changes in administrative organization. The legislation which followed was construed by many Tswana as an attack on the chiefship and the traditional order; in reply, the ruling Botswana Democratic Party asserted their commitment to the creation of a modern development-orientated regime in which patterns of the past were respected but not uncritically retained.

Essentially, the new system of local government is based upon three major laws. The first, the *Chieftainship Law, 1965*, regulates the recognition, suspension or removal of chiefs and their representatives, deputy chiefs and headmen. While the designation of chiefs remains an indigenous prerogative, the president maintains final sanction over the outcomes of local level political processes. The appointment of subordinate office-holders generally requires that the chief discuss proposals with the local population and then forwards them to the Ministry of Local Government and Lands.

The second, the *Local Government (District Councils) Law, 1965*, established district councils elected by popular vote along party political lines.[15] Countrol over matters of local policy passed to the council and its sub-committees, with administrative and financial functions being managed by a modernized bureaucracy. Council

---

[13] The *Bantu Homelands Citizenship Act* of 1970.
[14] In a well-known example of this, the Ngwaketse chief, Bathoen, left his office to become leader of the opposition Botswana National Front in 1970.
[15] The composition and electoral arrangements of the nine councils differs in accordance with local contingencies.

chairmen are elected by council members; and, although most of the early ones were chiefs, few now are.

The transfer of control away from traditional authorities, who had enjoyed greater autonomy than their counterparts in most other British territories, was justified in a White Paper (No. 21 of 1964): "... it cannot be expected that the chief, however enlightened and hard-working, will win popular support and active co-operation in carrying out policies which are ultimately not of his own and his people's making ... Local government therefore should not depend on the chiefs." Consistent with this policy, many important resources were also removed from chiefly hands. Thus the distribution of land became the exclusive function of Land Boards created by the *Tribal Land Act, 1968*. This third Act formalized the procedure for the allocation, registration and arbitration of land claims, and set up the necessary clerical and administrative machinery. This legislation intends that *all* immovable rural holdings will eventually be registered in order to reduce disputes and increase the efficacy of land use planning. A start has been made in this direction, but progress has been slow. The Land Board itself is ultimately controlled by the District Council and the Ministry of Local Government and Lands, each of which designates two of its members. The chief is an *ex officio* board member and has the right to appoint another. Initially, chairmen were usually chiefs; but, since 1974–5, the Ministry has begun to select others.

Similarly, other legislation[16] has transferred the potentially valuable rights over stray cattle from the chief to the district administration. The latter now manages the collection of most local finance (local tax, court fines, licence fees, etc.) as well as its distribution in the form of salaries, funds for development, welfare, education and so on. Moreover, the maintenance of order has become the task of the Botswana police and council-paid local policemen. The control of force is no longer the domain of office-holders and their regiments.

Finally, the district administration has assumed direct responsibility for supervising local development through its District Development Committee (D.D.C.). Ideally, each village should have a Village Development Committee (V.D.C.) to organize new projects with the (council) Assistant Community Development Officer.[17] The projects—many of which involve council finance—are then forwarded to the D.D.C. for approval. The achievements of the V.D.C.s have not been noteworthy to date. They have often been opposed vigorously by chiefs, without whose aid their success is inevitably limited. In the past, public projects were a chiefly concern; indeed, the success of office-holders depended partly upon material innovation. Many Tswana suggest that the modern chief is essentially a civil servant. His " subjects " are represented in parliament and district council by elected representatives, and his effective power is a function of personal factors rather than of his formal authority. The only aspect of the office which is relatively unaltered is the judicial one. The *Customary Courts Proclamation, 1961*, under which Chief's Courts are recognized and warranted, remains in force.[18]

In contrast to the South African regime, the Botswana government has tried to develop an administrative system which pays only limited respect to traditional institutional forms. The fact that many of the districts correspond largely with the territories of the individual chiefdoms, and that the council headquarters are usually located in their capitals, serve as reminders of this to rural populations. Across the border, the " encouragement of tribal traditions and institutions " is invoked to

---

[16] The *Matimela Act*, 1966.

[17] For a description of early community development in Botswana, see Peter Wass. "A case history: community development gets established in Botswana". *Int. Rev. community Dev.*, 16, 87/90, pp. 181–98, 1969.

[18] Although there have been minor changes in the extent of sentences and settlements which the Customary Courts may impose (this being stipulated for each one in its warrant), the nature of jurisdiction remains much as it is described on pp. 55–6 above.

justify the homeland system; in Botswana, the concept of "tribalism" has come to have explicitly negative and reactionary connotations. But, despite great differences of ideology and rhetoric, some of the social transformations undergone by the Tswana in the two countries bear strong similarities.

Tswana Transformations

Two themes emerge clearly from recent studies: first, the structure of Tswana chiefdoms (i.e., the hierarchy of agnatically based co-residential units) seems remarkably persistent in the face of dramatic external change; but, second, this structure is substantially transformed when chiefly power is diminished and authority relations altered.[19] This is related to the fact that indigenous residential arrangements exist *in tension* with ecological factors: given rainfall patterns and the contingencies of dry land crop cultivation and stock management, it would be in the material interests of tribesmen to live scattered alongside their agricultural holdings rather than in compact settlements. The need to move out from village to fields immeasurably complicates the organization of production. Indeed, among the Barolong, the chiefdom with the highest rates of arable production in Botswana, the population *does* live permanently scattered. Concentrated settlement has been resisted there expressly because this would decrease outputs. In other words, the general residence pattern occurs *in spite of* ecological and material considerations. The reasons for this lie in the fact that the chiefship is village based.

It is in the interests of the chief that tribesmen be domiciled in the capital, since his political control is most effectively exercised when they are concentrated. As a result, office-holders endeavour to exert pressure upon their subjects to return to the village whenever possible.[20] Similarly, it is generally the concern of headmen, and others who wish to gain influence, that people should not scatter permanently. For, while Tswana may move between two or more loci of activity, the major public political arena is firmly established at the centre.

The political factor was particularly persuasive in the past, then, because village concentration was actively encouraged by those who controlled such key resources as land, stock and the devolution of political offices. As the chiefship and established patterns of authority weaken, and key resources are placed in other hands, the centrifugal tendency increases. Families begin to leave the village and live in scattered households. No longer do political considerations determine people's actions and, slowly, the pattern of village concentration may disappear. In practice, the process is not quite so simple, for other factors may intervene (see below); nevertheless, this is its essential logic.

This logic has further implications. The enduring groupings of Tswana society (family groupings, wards, etc.) are formed in the village, and take on their politico-administrative and structural identity in that context. Hence the breakdown of villages could lead to more general structural dissolution: once the basic principles of recruitment and social organization cease to operate, other fundamental transformations must follow. In short, critical changes in the indigenous political system, and the chiefship at its centre, may set in motion a complex process of related changes. It is clear, therefore, that government policies in respect of the chiefship must have crucial effects upon other aspects of indigenous life. Given these considerations, it is necessary to examine whether such processes have occurred to date.

(a) Government and politics

I have already stressed that South Africa and Botswana established their respective systems upon opposed premises concerning traditional institutions. In Botswana,

[19] Kuper, 1975 (a): 144–7, summarizes the evidence.
[20] In most chiefdoms it is theoretically illegal to be domiciled outside a village. But the ability to enforce this depends upon the power of individual chiefs.

the explicit reduction of chiefly authority has meant that, while office-holders may retain their advisers and councils, and perpetuate the *forms* of indigenous government (p. 52 above), much of its substantive *content* is lost. Chiefly councils and public gatherings are primarily contexts for the dissemination, discussion and planning of administrative matters at the behest of higher authorities; their significance as loci of internal *political* processes appears to be diminishing rapidly. Thus, although the judicial functions of the chiefship remain intact, and its incumbents act as liaisons between government and the rural population, the centrality of the office has been significantly undermined.

Their loss of constituted authority does not necessarily mean that chiefs and headmen no longer have *any* access to power. But the extent to which they exercise influence depends largely upon personal skills. By exploiting the respect which some of the population retain for the office,[21] a perceptive traditional leader may extend his legitimacy. Moreover, a number of residual politico-economic resources may remain at his disposal. For example, a recent essay[22] has shown that an able office-holder may manipulate his intercalary position between the administration and his people in order to recruit support for himself from both sources. Membership of the District Council, the Land Board, perhaps the V.D.C., and other bodies may enable the active local authority to have a considerable effect upon public affairs. In addition, their former hegemony over economic resources has left some chiefs and headmen in control of considerable personal wealth in land, cash and other capital holdings (boreholes, agricultural machinery, transport facilities, etc.). These assets may be employed for both economic and political purposes; their scarcity guarantees high returns, in terms of both cash and patronage, from loans and the extension of usufruct. By virtue of their activities in this sphere, a chief or headman may expand his informal control. As this suggests, the *de facto* power of modern office-holders is highly variable, and is achieved in spite of severe limitations upon authority. Of course, such variability existed in the past as well,[23] but for rather different reasons.

The circumstances which have transformed the chiefship have also created opportunities for the emergence of other powerful individuals. Hence wealthy commoners (merchants, successful farmers, agricultural entrepreneurs) and recruits to the cadres of government (M.P.s, party leaders, district councillors, government employees) may figure prominently in the affairs of their local communities. Again, their influence is not uniform: unfortunately, however, the political role of " new men " in rural Botswana is not yet systematically documented.

The South African situation appears more complex. As in Botswana, the chief and his close agnates *nominally* remain the central figures at the local level. Here, too, the external forms of traditional government remain and are, in fact, more comprehensively integrated into the modern administrative hierarchy than is the case across the border.[24] Furthermore, a number of functions which have passed to the district authorities in Botswana remain in the hands of tribal authorities in Bophuthatswana (e.g. land allocation, the administration of certain funds, etc.). Officials of both governments seek co-operation from chiefs and headmen whenever they require meetings to discuss (or propagate) new policy and arrangements; of the two, however, the South Africans regard the opinions of the traditional authorities more seriously and treat them as symptomatic of public attitudes. As a result, Bophuthatswana

---

[21] While the Tswana maintained a strong respect for the chief*ship*, the performance of office-holders was always subjected to the critical evaluation of tribesmen. Chiefs had to *earn* respect by ruling well. Hence, respect for the office never assured its holder of power *per se*; it was a resource which required exploitation. See J. L. Comaroff, 1975: *passim*.

[22] A. Kuper, "Gluckman's Village Headman", *American Anthropologist*, 72(2): 355–358, 1970.

[23] See note 21 above.

[24] These factors have persuaded Moloto, 1974, Jeppe, 1974 and Breutz, 1958, to suggest that traditional government persists in all but matters of superficial detail in modern South Africa.

chiefs and headmen are lobbied directly when public response is held to be of political consequence. But these factors are *not* an index of the extent to which traditional institutions persist. Rather, they simply indicate the degree to which any government exploits the formal features of such institutions in the management of political control and administrative order.

A recent study of chiefship in Bophuthatswana[25] suggests that, notwithstanding stated policy, "separate development" has had a profoundly disruptive effect upon the indigenous political systems. For, in integrating so-called traditional institutions into the homeland hierarchy, the South African government has treated these Tswana systems as if they were solely *administrative* ones, the function of which lay simply in the management of order and public affairs. Little attention was paid to the fact that indigenous government was itself located in a *political* system, or that the chiefship provided the focus for ongoing political competition. The transformation of tribal government into an administrative organ of the state has turned the tribal authorities into a closely supervised echelon of lower level bureaucrats. No longer are they the politicians in a dynamic system. The resources formerly employed by chiefs and headmen in order to exercise power and increase legitimacy have passed largely into other hands. As in Botswana, able individuals may still gain some personal influence; but the essence of Tswana politics has been effectively eliminated.

The introduction of the new arrangements, then, has had broadly similar effects in both countries: the external forms of the indigenous system remain, and have been incorporated—to a greater or lesser extent—into the emergent administrative hierarchies. The political content of these systems has been largely excised, and the substance of public affairs is regulated by the respective governments. As a result, the chiefship, which constituted the institutional focus for the negotiation of power relations and competition for influence, has been largely disassociated from the political and economic resources which underpinned its centrality in Tswana life. Some office-holders manage to retain personal followings and contrive to manipulate residual resources. But the office is not a substantial source of legitimacy today, and power must be shared with "new men" whose position derives from the exploitation of the modern politico-economic environment.

(b) Social organization

Given the functional relationship between the Tswana political order and patterns of social organization (p. 71 above), changes wrought in the former by government action should be accompanied by major social transformations. Of course, the innovations are very recent—in some communities they have barely been felt as yet—but the trends seem clear. Thus, for example, the South African government weakened the Tlhaping chiefship earlier than most in Bophuthatswana, and the office-holder duly relinquished control over most crucial resources. The population soon deserted the capital and now lives scattered throughout the Taung Reserve. The administrative units which have replaced the former wards are defined by geographical propinquity alone. Households are situated alongside fields and cattle-posts, "according to no particular pattern."[26] The basic organizing principles of Tswana structure no longer operate here: there are no residentially contiguous family groups of agnatically related households, no wards or large settlements, very low incidences of patrilocal residence or close kin marriage. The chiefdom has been reduced to an undifferentiated collectivity of homesteads. Similar processes occurred in pre-independence Botswana among the Tawana[27] and the Khurutse,[28] who lived under the control of a concession company. In both situations, the weakening of the chiefship was followed by a dispersion similar to that of the Tlhaping. Only when the centrality of the office

[25] Comaroff. 1974: 36–51.
[26] Pauw, 1960(b): 52.
[27] Ashton, 1937.
[28] Werbner, 1971.

was restored did the population return to live in wards within concentrated settlements. Today, it seems, the chiefship could not be re-established so easily.

The Tlhaping example is an extreme one and the likelihood of similar outcomes is difficult to ascertain on the basis of the available data. But it is clear that this is one direction which the process of transformation may take. A second is exemplified by the Tshidi-Rolong of the Mafeking district. Here there has also been a progressive weakening of the chiefship since 1956.[29] The social correlates, however, have been different to those observed among the Tlhaping. A number of families have left the capital and become domiciled at extra-settlement holdings. But the majority have not done so for expressly material reasons: the adjoining white town is an administrative and railway centre which offers employment opportunities and certain public facilities. In fact, some Tshidi-Rolong who formerly lived in more remote villages have immigrated into the capital as a result of the impoverishment of their areas.

Superficially, the structure and physical form of the village appear to persist here. But its sociological character is changing. Wards are steadily ceasing to function as politico-administrative groupings and are becoming mere territorial neighbourhoods, similar to the administrative units of the Tlhaping. The moribund ward has a number of typical political, social and economic features: the headman ceases to participate in the government of the chiefdom and does not attend the chiefly council *(lekgotla)* or act as a liaison between his grouping and the central authorities; he does not call meetings or hear cases and exercises no control over immigration into, or emigration from, the unit; if he is a migrant labourer, he does not appoint a representative to act in his stead and, when he dies, a successor may not be designated; the ward tends to lose its agnatic core, partly because patrilocal residence gives way to neo-local patterns; similarly, the component family groups may begin to lose their cohesion, this being reflected in the fact that life crisis situations (marriage, birth and death) are managed increasingly by the individual household;[30] there are marked rises in the incidences of casual unions and illegitimacy; given the absence of a large number of men as migrant labourers, this produces increasing numbers of domestic groupings comprising women, their unmarried children and, perhaps, illegitimate grandchildren; internal exchange relations slowly assume a contractual and competitive character and, beyond the household, reciprocal labour, credit and service arrangements disappear; many cease to engage in agriculture, or to move annually to their holdings, and depend upon the labour market and petty trade for their incomes; finally, members no longer take an active interest in the public affairs of the chiefdom or the ward, and the latter soon loses its traditional ideology.

These characteristics tend to manifest themselves gradually and cumulatively. The process is not irreversible but, among the Tshidi-Rolong, there has been a constant increase in the number of moribund wards.

In summary, then, it seems that the process of change may take one of two directions, exemplified in their extreme forms by the Tlhaping and the Tshidi-Rolong respectively. These may be typified as follows:

(i) *The Tlhaping-type:* the population disperses and isolated household units are permanently established alongside fields and cattle-posts. There is no longer any large settlement, or correspondence between ties of kinship and residential propinquity. Co-operation beyond the domestic unit tends to give way to contractual relations and commercial transactions. Depending upon circumstance, administrative units may be introduced; but they will be defined entirely in terms of geographical convenience and will lack any kinship basis or political identity.

---

[29] Comaroff, 1974: 36–51.
[30] Some of these social features are, of course, associated directly with the disruptive effects of labour migration. See below, p. 75f.

It is patent that the erosion of the indigenous political systems has *not* led to the universal fragmentation of villages; and, in many situations, the trends (to the degree that they may be observed as yet) suggest that it will not do so. This is due to the intervention of a number of factors, of which two are the most compelling: the relative availability of economic opportunities and public facilities within or nearby the capital village; and the regulation of mobility by agencies of government (the T.L.A. in South Africa) or bodies created by them (Land Boards in Botswana).[31] While the absence of these factors is likely to be associated with "Tlhaping-type" processes, their presence may create suitable conditions for the second "type":[32]

(ii) *The Tshidi-Rolong-type:* a proportion of the population may leave the village, but the majority remain in it. Gradually, however, wards become moribund as politico-administrative groupings. They are reduced to neighbourhood living-quarters composed of largely unrelated households, and lose their internal structural form. As in the "Tlhaping-type," there is a growing atomization of the domestic grouping. In becoming an isolated social and economic unit, its ties with other such units (including those with whom kinship linkages may exist) are also defined increasingly in contractual, and less in co-operative terms. Agnatic cores, patrilocal recruitment and cousin marriage cease to be characteristic of most social units.[33]

Underlying both "types," of course, is the fundamental principle that Tswana social forms are an epiphenomenon of their political processes.[34] While the chiefship remains the fulcrum of a dynamic *political* system, these processes serve to articulate the ward (and other social units) into the political community.[35] When the chiefdom is reduced to an administrative organ, however, its members and component groupings merge into an undifferentiated collectivity. In response to contrasting exogenous factors, the Tlhaping and Tshidi-Rolong have chosen alternative geographical loci in which to express this. But, ultimately, the logic which informs both transformations is the same.

Conclusion

Although I have attempted to identify and account for dominant trends of social transformation among rural Tswana, it has not been possible to catalogue the enormous number of small-scale changes which have occurred since 1953. Nor do I suggest that all such changes are directly related to the processes described above; for it is not *only* upon the transformation of the indigenous political system that they are contingent. Hence, in conclusion, three further points must be made.

First, it would appear that most of the substantive changes noted in the recent literature—beyond those discussed above—are, in themselves, ones of frequency and magnitude rather than of kind. Thus, for example, in the sphere of religion (pp. 58–66 above), nominal adherence to Christianity continues to grow, the ongoing proliferation of small sects produces increasing numbers of churches, and emergent patterns of syncretistic belief and ritual undergoes gradual accretion and elaboration.[36] But these processes began in the nineteenth century and, while modern conditions

---

[31] In Botswana, this reflects a government policy which encourages village concentration in order to expedite the extension of health, education, welfare and agricultural facilities.

[32] Under exceptional circumstances (e.g. the Botswana Rolong), commercial farming opportunities may provide incentives which are sufficiently strong to counter the intervening factors, even where they do exist.

[33] In broad sociological character, if not in historical process, this represents a change from African towns of Southall's type A to type B. See A. Southall, pp. 6–13 in Southall (ed.) *Social Change in Modern Africa* (1961).

[34] Kuper, 1975(a): *passim.*

[35] Comaroff, 1973(a): especially chapters III, VIII.

[36] Jean Comaroff, 1974: *passim.*

(especially in South Africa) may add to the attraction of the churches, the sociological character of Tswana religion has changed little during the period under review. Similarly, the rate of labour migration (pp. 30–1) is constantly rising, and its control is extended ever more rigidly. Yet the disruptive consequences of this system itself for rural social forms are not significantly different from those which were manifest before 1953. They simply concern more people in a more pervasive fashion. This is not to say that the relationship between the system of labour migration and the South African political economy has not altered with the elaboration of *apartheid*; for those affected, however, it generates broadly the same social problems as it did prior to the accession of the Nationalist Party. But this, too, is changing.

Second, then, is the fact that several such changes which were initiated prior to 1953 by external factors, like rural-urban mobility, begin to take on a new dimension in the context of the transformational processes which I have outlined. For instance, the spiralling incidence of casual unions, illegitimacy and domestic units lacking adult males is not a recent occurrence in many chiefdoms. While wards and their component family groups continue to exist, the groupings and individuals concerned may be integrated into larger units, and into their productive and co-operative activities; indeed, arrangements for guardianship over people of this category are implicit in the rules associated with agnatic relations.[37] But, once either the "Tlhaping-type" or the "Tshidi Rolong-type" of process begins, such integration is weakened, and unmarried motherhood, widowhood and other forms of social isolation assume a new meaning; they may, in turn, necessitate new patterns of social and economic adaptation.

Finally, a significant area of change has arisen from the extension of new economic opportunities into the rural areas. In Botswana, this is most notable in the agricultural sector, in which the relevant Ministry has introduced a number of schemes aimed at raising the productivity and profitability of arable and pastoral enterprises.[38] The National Development Bank has supported such development by providing loan capital under favourable conditions. This has led to the emergence of a small number of wealthy commercial farmers, some of whom have also gained control of informal trade in the hire of agricultural machinery, of petty grain speculation, and of rural trading stores. For a few of these men, success in the country has led to a career in government and politics, or to financial expansion into the urban centres; others have remained in their chiefdoms and have become prominent in local and district affairs (see above, p. 72). In South Africa, the Bophuthatswana administration constantly expresses concern over rural development; but it seems that a significant proportion of successful Tswana depend directly upon entrepreneurial rather than agricultural activities, some having been financed by the state controlled Bantu Investment Corporation.[39] Again, people of this category tend to become influential in their local communities. But a political or financial career of broader scope is, of course, largely closed to them. In both countries, the role of these " new men " may provide the key to the future of rural politics; systematic research into it, however, is just beginning.

[37] Roberts, 1972, Schapera, 1938.
[38] For a description of agricultural extension in Botswana, see B. G. Lever. *Agricultural extension in Botswana*. Univ. of Reading Dept. of Agricultural Economics, Development Study no. 7, 1970.
[39] The senior ranking official of the B.I.C. at Mafeking suggested this to me in an interview in 1970.

SELECT BIBLIOGRAPHY

The works listed below include all those cited in the text, together with some others of importance. More detailed bibliographies of the Tswana are available in the following:

Doke, C. M.
    1933. A Preliminary Investigation of the State of the Native Languages of South Africa. *Bantu Studies*, **7**, pp. 1–98. (Bibliography of Tswana language and literature, compiled by G. P. Lestrade, pp. 77–85.)

Holden, M. A., and Jacoby, Annette
    1950. *A Select Bibliography of South African Native Life and Problems. Modern Status and Conditions*, 1939–1949. (Supplement to Schapera, 1941.) University of Cape Town, School of Librarianship. (Tswana, Part I, p. 9; Part II, pp. **12** f., and *passim*.)

Lestrade, G. P.
    1944. Some recent publications concerning languages of the Sotho group. *African Studies*, **3**, pp. 22–27.

Letele, G. L.
    1944. Some recent literary publications in languages of the Sotho group. *African Studies*, **3**, pp. 161–71.

Schapera, I. (ed.)
    1934. The present state and future development of ethnographical research in South Africa. *Bantu Studies*, **8**, pp. 219–342. (Tswana, pp. 244–6, 319–22.)
    1941. *Select Bibliography of South African Native Life and Problems.* London: Oxford University Press. (Tswana ethonography, pp. 83–8; modern status and conditions, pp. 116, 138–41, 151, 171 f., 181 f.; language and literature, pp. 227–30.)

Stevens, Pamela
    1947. *Bibliography of Bechuanaland.* University of Cape Town: School of Librarianship.

GENERAL

Agar-Hamilton, J. A. I.
    1937. *The Road to the North: South Africa*, 1852–1886. London: Longmans. (Political history.)

Ashton, E. H.
    1937. Notes on the political and judicial organization of the Tawana. *Bantu Studies*, **11**, pp. 67–83.
    1947. Democracy and indirect rule. *Africa*, **17**, pp. 235–51.
    1949. The High Commission Territories. *In* Hellmann and Abrahams (eds.), pp. 706–41.

Barnes, L.
    1932. *The New Boer War.* London: Hogarth Press. (Review of conditions in the B.P., pp. 128–201.)

Bechuanaland Protectorate
    1924—. *Minutes of the .... Native Advisory Council.* (In progress; annual.) (Minutes for 1920–3 not published.)
    1937. *Report of the Commissioner appointed .... to advise on Medical Administration in Bechuanaland Protectorate.* Mafeking: Principal Medical Officer's Office.
    1949. *The Laws of the Bechuanaland Protectorate ....* Revised edition .... prepared by H. C. Juta. 3 vols. London: Crown Agents for the Colonies.
    1950—. *Minutes of the Joint Advisory Council.* (In progress; annual.)
    1952. *Census, 1946.* (Mafeking: Controller of Stores.)

Bent, J. T.
    1892. Among the chiefs of Bechuanaland. *Fortnightly Rev.*, **51**, pp. 642–54.

Breutz, P. L.
    1941. *Die politischen und gesellschaftlichen Verhältnisse der Sotho-Tswana in Transvaal und Betschuanaland.* Hamburg: Friedrichsen.

Broadbent, S.
    1865. *A narrative of the first introduction of Christianity amongst the Barolong tribe of Bechuanas, South Africa.* London: Wesleyan Mission House.

Brown, J. T.
    1921. Circumcision rites of the Becwana tribes. *J. R. Anthrop. Inst.*, **51**, pp. 419–27.
    1926. *Among the Bantu nomads: a record of forty years spent among the Bechuana.* London: Seeley Service.

Burchell, W. J.
    1822–4. *Travels in the interior of South Africa.* 2 vols. London: Longmans. (Tlhaping.)

Campbell, J.
    1813. *Travels in South Africa.* London: Black, Parry. (Tlhaping.)
    1822. *Travels in South Africa ....*, *being a narrative of a second journey.* 2 vols. London: Westley. (Southern tribes, especially Tlhaping and Hurutshe.)

Chamberlin, D. (ed.)
    1940. *Some letters from Livingstone, 1840–72.* London: Oxford University Press. (Good source book for history of Kwena, Ngwato, and others.)

Chapman, J.
    1868. *Travels in the interior of South Africa.* 2 vols. London: Bell & Daldy. (Ngwato and other northern tribes.)

Couperthwaite, B.
    1951. The Bechuanaland Protectorate. *Race Relations J.*, **18**, pp. 27–71. (Review of current conditions.)

Debenham, F.
    1948. *Report on the water resources of the Bechuanaland Protectorate, Northern Rhodesia,* (etc.). London: H.M.S.O., Colonial Research Publications, No. 2. (Bechuanaland, pp. 31–38.)

Dornan, S. S.
    1925. *Pygmies and Bushmen of the Kalahari.* London: Seeley Service. (Tswana, pp. 243–312.)

du Plessis, J.
    1911. *A history of Christian missions in South Africa.* London: Longmans. (Important general survey.)

Ellenberger, V. F.
    1937. History of the Ba-ga-Malete of Ramoutsa (Bechuanaland Protectorate). *Trans. roy. Soc. S. Afr.*, **25**, pp. 1–72.
    1939. History of the BaTlôkwa of Gaberones (Bechuanaland Protectorate). *Bantu Studies*, **13**, pp. 165–98.

Evans, I. L.
    1934. *Native policy in Southern Africa.* Cambridge: University Press. (Bechuanaland, pp. 82–93.)

Frazer, J. G.
    1910. *Totemism and Exogamy.* 4 vols. London: Macmillan. (Summary of Tswana data, Vol. II, pp. 369–77.)

Fritsch, G.
    1868. *Drei Jahre in Süd-Afrika.* Breslau: F. Hirt. (Bechuanaland, pp. 274–399.)
    1872. *Die Eingeborenen Süd-Afrika's.* Breslau: F. Hirt. (Tswana, pp. 149–210.)

Great Britain: Colonial Office
    1886. *Report of the commissioners appointed to determine lands claims .... in British Bechuanaland.* London: H.M.S.O., C.4889.
    1896. *Correspondence re visit to this country of the chiefs Khama, Sebele, and Bathoen; and the future of the Bechuanaland Protectorate.* London: H.M.S.O., C.7962.
    1931. *Papers relating to the health and progress of native populations in certain parts of the Empire.* London: H.M.S.O., Colonial No. 65. (Bechuanaland, etc., pp. 144–52.)
    1952. *An economic survey of the Colonial territories, 1951. Vol. I. The Central African and High Commission Territories.* London: H.M.S.O., Colonial No. 281(1). (Bechuanaland Protectorate, pp. 75–88.)

Great Britain: Commonwealth Relations Office
    1898—. *Bechuanaland Protectorate: Annual Reports.* London: H.M.S.O. (In progress; formerly issued by Colonial Office.)
    1950. *Bechuanaland Protectorate: succession to the chieftainship of the Bamangwato tribe.* London: H.M.S.O., Cmd. 7913.

Great Britain: War Office
    1905. *The native tribes of the Transvaal.* London: H.M.S.O. (History and political conditions.)

Haccius, G.
    1907–20. *Hannoversche Missionsgeschichte.* 3 vols. Hermannsburg: Missionsbuchhandlung. (Valuable historical source for Transvaal and southern Protectorate tribes.)

Hailey, *Lord*
    1938. *An African survey: a study of problems arising in Africa south of the Sahara.* London: Oxford University Press.

Hellmann, Ellen, and Abrahams, Leah (eds.)
    1949. *Handbook of Race Relations in South Africa.* Cape Town: Oxford University Press.

Hepburn, J. D.
    1895. *Twenty years in Khama's country.* London: Hodder & Stoughton.

Hirschberg, W.
    1936. *Völkerkundliche Ergebnisse der südafrikanischen Reisen Rudolf Pöch's in den Jahren 1907 bis 1909.* Wien: Anthropologische Gesellschaft. (Material culture of the Tawana, pp. 26–30.)

Hodgson, Margaret L., and Ballinger, W. G.
    1932. *Britain in South Africa: (No. 2) Bechuanaland Protectorate.* Alice: Lovedale Press. (Critical account of British policy.)

Hole, H. M.
    1932. *The Passing of the Black Kings.* London: P. Allan. (Political history of Kgama and other chiefs.)

Holub, E.
    1881. *Sieben Jahre in Süd-Afrika.* 2 vols. Wien: A. Hölder. (Kwena, Ngwato.)

Jennings, A. E.
    1933. *Bogadi: a study of the marriage laws and customs of the Bechuana tribes of South Africa.* Tiger Kloof: London Missionary Society. (Propaganda pamphlet.)

Joyce, J. W.
    1938. Report on the Masarwa in the Bamangwato Reserve, Bechuanaland Protectorate. *League of Nations Publications, VI.B. Slavery* (C.112.M.98. 1938. VI), Annex 6, pp. 57–76.

Khama, Tshekedi
    1936. Chieftainship under indirect rule. *J.R. Afr. Soc.*, **25**, pp. 251–61.

Kirby, P. R. (ed.)
    1939–40. *The Diary of Dr. Andrew Smith, 1834–6.* Cape Town: van Riebeeck Society; Publications Nos. 20, 21. (Important source book on Tlhaping and other southern tribes.)

Kohler, J.
    1902. Das Recht der Betschuanen. *Z. vergl. Rechtswiss.*, **15**, pp. 321–36.

Kotzé, D. J. (ed.)
    1950. *Letters of the American Missionaries, 1835–8.* Cape Town: van Riebeeck Society; Publication No. 31. (Hurutshe.)

Krüger, F.
    1935. Das Recht der Sotho-Chuana-Gruppe der Bantu in Südafrika. *Mitt. Sem. Orient. Spr.* (Berlin), **38** (Abt. III), pp. 53–144. (Compilation; bibliography.)

Kuczynski, R. R.
    1949. *A demographic survey of the British Colonial Empire. Vol. II: South African High Commission Territories, East Africa* (etc.). London: Oxford University Press.

Language, F. J.
    1941. *Kapteinskap onder die Tlhaping.* Unpublished Ph.D. thesis, University of Stellenbosch. (Some chapters, but not the whole work, have been separately published, as follows:)
    1942. Herkoms en Geskiedenis van die Tlhaping. *African Studies*, **1**, pp. 115–33.
    1943 (a). *Stamregering by die Tlhaping.* Stellenbosch: Pro Ecclesia-Drukkery.
    (b) Die verkryging en verlies van lidmaatskap tot die stam by die Tlhaping. *African Studies*, **2**, pp. 77–92.
    (c) Die bogwêra van die Tlhaping. *Tydskrif vir Wetenskap en Kuns*, **4**, pp. 110–34.

Lebzelter, V.
    1933. Das Betschuanendorf Epukiro (Südwestafrika). *Z. Ethn.*, **65**, pp. 44–74. (Tlharo offshoot.)

Lemue, P.
    1844 (a) Coutumes religieuses et civiles des Béchuanas. *J. Missions évangél.*, **19**, pp. 49–58.
    (b) Nouveaux détails sur les superstitions des Béchuanas. *J. Missions évangél.*, **19**, pp. 401–10.
    1854. La circoncision chez les Africains du sud. *J. Missions évangél.*, **29**, pp. 208–13. (Tlhaping and Rolong.)

Lestrade, G. P.
    1926. Some notes on the *bogadi* system of the BaHurutshe. *S. Afr. J. Sci.*, **23**, pp. 937–42.
    1928. Some notes on the political organization of the BeChwana. *S. Afr. J. Sci.*, **25**, pp. 427–32. (Hurutshe.)
    1929. The Suto-Chuana tribes, and the Bechuana. In *The Bantu Tribes of South Africa: reproductions of photographic studies by A. M. Duggan-Cronin* (Cambridge: Deighton Bell), Vol. II, section 1, pp. 7–23.

Lichtenstein, M. H. C.
    1811–12. *Reisen im südlichen Afrika in den Jahren 1803–1806.* Berlin: Salfeld. (Tlhaping.)

Lister, Margaret H. (ed.)
    1949. *Journals of Andrew Geddes Bain.* Cape Town: van Riebeeck Society; Publication No. 30. (Rolong and Ngwaketse in 1826.)

Livingstone, D.
 1857. *Missionary travels and researches in South Africa.* London: Murray. (Kwena and southern tribes.)

Lloyd, E.
 1895. *Three great African chiefs: Khâmé, Sebelé, and Bathoeng.* London: T. Fisher Unwin. (Popular history.)

London Missionary Society
 1935. *The Masarwa (Bushmen): report of an inquiry.* Tiger Kloof: L.M.S. Book Room. (Rejoinder to Tagart's criticisms of Ngwato policy.)

Lovett, R.
 1899. *The history of the London Missionary Society, 1795–1895.* 2 vols. London: Oxford University Press. (South Africa, including Bechuanaland: Vol. I pp. 481–670.)

Mackenzie, J.
 1871. *Ten Years north of the Orange River.* Edinburgh: Edmonston & Douglas. (Important source book on Ngwato history and culture.)
 1883. *Day-dawn in dark places: a story of wanderings and work in Bechwanaland.* London: Cassell. (Popular sequel to 1871.)
 1887. *Austral Africa: losing it or ruling it.* 2 vols. London: Sampson Low. (Political history dealing mainly with extension of European control.)

Matthews, Z. K.
 1940. Marriage customs among the Barolong. *Africa*, **13**, pp. 1–24.
 1945. A short history of the Tshidi Barolong. *Fort Hare Papers*, **1**, pp. 9–28.

Mockford, J.
 1931. *Khama: King of the Bamangwato.* London: Cape. (Popular biography.)
 1950. *Seretse Khama and the Bamangwato.* London: Staples Press. (Same as 1931, with some additional material.)

Moffat, R.
 1842. *Missionary Labours and Scenes in Southern Africa.* London: Snow. (Tlhaping.) (See also: Schapera (ed.), 1951 ; Wallis.)

Molema, S. M.
 1920. *The Bantu, Past and Present.* Edinburgh: Green. (History, culture, and modern conditions, by a Rolong medical practitioner.)
 (1951). *Chief Moroka: his life, his times, his country, and his people.* Cape Town: Methodist Publishing House. (History of Seleka-Rolong.)

Nettelton, G. E.
 1934. History of the Ngamiland tribes up to 1926. *Bantu Studies*, **8**, pp. 343–60. (Tawana.)

Newton, A. P., and Benians, E. A. (eds.)
 1936. *The Cambridge History of the British Empire. Vol. VIII: South Africa, Rhodesia, and the Protectorates.* Cambridge: University Press.

Norton, W. A.
 1922. Circumcision regiments as a Native chronology. *Trans. roy. Soc. S. Afr.*, **10**, pp. 245–51.

Passarge, S.
 1905. Das Okawangosumpfland und seine Bewohner. *Z. Ethn.*, **37**, pp. 649–716. (Tawana.)
 1908. *Südafrika: eines Landes-, Volks-, und Wirtschaftskunde.* Leipzig: Quelle & Meyer.

Pim, A. W.
 1933. *Financial and Economic Position of the Bechuanaland Protectorate.* Report of the Commission appointed by the Secretary of State for Dominion Affairs. London: H.M.S.O., Cmd. 4368.

Pole Evans, I. B.
 1948. *A reconnaissance trip through the Eastern portion of the Bechuanaland Protectorate, April, 1931, and an expedition to Ngamiland, June–July, 1937.* Pretoria: Government Printer. Botanical Survey of S. Africa: Memoir No. 21.

Richards, Audrey I.
 1941. Some causes of a revival of tribalism in South African Native Reserves. *Man*, **41**, pp. 89–90. (Mosêtlha-Kgatla.)

Rogers, H., and Linington, P. A.
 1949. *Native administration in the Union of South Africa.* Pretoria: Government Printer. (Official handbook.)

Rolland, S.
 1843. Idées religieuses et coutumes des Béchuanas. *J. Missions évangél.*, **18**, pp. 473–9. (Southern tribes.)

Sandilands, A.
 1939. A Sengwato wedding. *Tiger Kloof Magazine*, No. 21, pp. 16–29. (Modern ceremonies.)

Sargant, E. B.
    1906. *Report on education in the Bechuanaland Protectorate*, 1905. Unpublished MS. in Government archives, Mafeking.
    1908. *Report on Native education in South Africa. Part III: Education in the Protectorates.* London: Longmans. (Important discussion of policy and effects.)

Schapera, I.
    1930. (*a*) Some ethnographical texts in SeKgatla. *Bantu Studies*, **4**, pp. 73–93. (With translations and notes.)
        (*b*) The " little rain " (*pulanyana*) ceremony of the Bechuanaland BaKxatla. *Bantu Studies*, **4**, pp. 211–6.
    1933. (*a*) Premarital pregnancy and Native opinion: a note on social change. *Africa*, **6**, pp. 59–89. (Kgatla.)
        (*b*) The Native as letter-writer. *The Critic* (Cape Town), **2**, pp. 20–28. (Kgatla.)
        (*c*) Economic conditions in a Bechuanaland Native Reserve. *S. Afr. J. Sci.*, **30**, pp. 633–55. (Kgatla.)
    1934. (*a*) Oral sorcery among the natives of Bechuanaland. In *Essays presented to C. G. Seligman* (ed. E. E. Evans-Pritchard and others. London: Routledge), pp. 293–305. (Kgatla.)
        (*b*) Herding rites of the Bechuanaland BaKxatla. *Amer. Anthrop.*, **37**, pp. 561–84.
        (*c*) Present-day life in the Native reserves. In *Western Civilization and the natives of South Africa* (ed. I. Schapera. London: Routledge), pp. 39–62. (Kgatla.)
    1935. The social structure of the Tswana ward. *Bantu Studies*, **9**, pp. 203–24. (Kgatla and Ngwato.)
    1936. The contributions of Western civilization to modern Kxatla culture. *Trans. roy. Soc. S. Afr.*, **24**, pp. 221–52.
    1938. *A Handbook of Tswana Law and Custom.* London: International Institute of African Languages and Cultures.
    1940. (*a*) *Married Life in an African Tribe.* London: Faber. (Kgatla.)
        (*b*) The political organization of the Ngwato in Bechuanaland Protectorate. In *African political systems* (ed. M. Fortes and E. E. Evans-Pritchard. London: International Institute of African Languages and Cultures), pp. 36–82.
    1942. (*a*) A short history of the BaKgatla-bagaKgafêla of Bechuanaland Protectorate. University of Cape Town, School of African Studies: *Communications*, n.s. No. 3.
        (*b*) A short history of the BaNgwaketse. *African Studies*, **1**, pp. 1–26.
    1943. (*a*) *Native land tenure in the Bechuanaland Protectorate.* Alice: Lovedale Press.
        (*b*) The system of land tenure on the Barolong Farms. Unpublished report to the B.P. Government.
        (*c*) The land problem in the BaTlôkwa Reserve. Unpublished report to the B.P. Government.
        (*d*) The work of tribal courts in the Bechuanaland Protectorate. *African Studies*, **2**, pp. 27–40.
        (*e*) Tribal legislation among the Tswana of the Bechuanaland Protectorate. London School of Economics: Monographs on Social Anthropology, No. 9.
        (*f*) The Native land problem in the Tati district. Unpublished report to the B.P. Government. (Khurutshe.)
    1945. (*a*) The land problem in the BaMalete Reserve. Unpublished report to the B.P. Government.
        (*b*) Notes on the history of the Kaa. *African Studies*, **4**, pp. 109–21.
    1946. Some features in the social organization of the Tlôkwa (Bechuanaland Protectorate). *Southwestern J. Anthrop.*, **2**, pp. 16–47.
    1947. (*a*) *Migrant Labour and tribal life: a study of conditions in the Bechuanaland Protectorate.* London: Oxford University Press.
        (*b*) The political annals of a Tswana tribe. University of Cape Town, School of African Studies: Communications, n.s., No. 18. (Ngwaketse.)
    1949. The Tswana conception of incest. In *Social Structure: essays presented to A. R. Radcliffe-Brown* (ed. M. Fortes. Oxford: Clarendon Press), pp. 104–20.
    1950. Kinship and marriage among the Tswana. In *African Systems of Kinship and Marriage* (ed. A. R. Radcliffe-Brown and C. D. Forde. London: International African Institute), pp. 140–65.
    1952. (*a*) Sorcery and witchcraft in Bechuanaland. *African Affairs*, **51**, pp. 41–50.
        (*b*) *The ethnic composition of Tswana tribes.* London School of Economics: Monographs on Social Anthropology, No. 11.

Schapera, I. (ed.)
    1937. *The Bantu-speaking tribes of South Africa : an ethnographical survey.* London : Routledge.
    1951. *Apprenticeship at Kuruman : being the journals and letters of Robert and Mary Moffat,* 1820–8. London : Chatto & Windus. Central African Archives : Oppenheimer Series, No. 5.

Schapera, I., and van der Merwe, D. F.
    1945. *Notes on the tribal groupings, history, and customs, of the BaKgalagadi.* University of Cape Town, School of African Studies : Communications, n.s., No. 13.

Schultze, L.
    1907. *Aus Namaland und Kalahari.* Jena : G. Fischer. (Ngwaketse, pp. 620–49.)

Scroggie, Helen M. M.
    1946. *The sociology of Ngwaketse diet.* Unpublished M.A. thesis, University of South Africa.

Sillery, A.
    1952. *The Bechuanaland Protectorate.* Cape Town : Oxford University Press. (Political and tribal histories.)

Smith, E. W.
    1950. The idea of God among South African tribes. In *African Ideas of God* (ed. E. W. Smith. London : Edinburgh House Press), pp. 78–134.

South Africa
    1905. *South African Native Affairs Commission : Report and Minutes of Evidence.* 5 vols. Cape Town : Cape Times.

Squires, B. T.
    1943. Malnutrition amongst Tswana children. *African Studies,* **2,** pp. 210–4.
    1949. *The feeding and health of African school children : report on the Kanye nutrition experiment.* University of Cape Town, School of African Studies : Communications, n.s., No. 20. (Ngwaketse and Kwena.)

Stow, G. W.
    1905. *The native races of South Africa.* London : Sonnenschein. (Tswana totemism, history, and traditions, pp. 404–561.)

Tagart, E. S. B.
    1933. Report on the conditions existing among the Masarwa in the Bamangwato Reserve of the Bechuanaland Protectorate. *Official Gazette of the High Commissioner for South Africa,* Vol. 122, No. 1,662 (12 May).

Transvaal : Native Affairs Department
    1905. *Short history of the Native tribes of the Transvaal.* Pretoria : Government Printer.

Union of South Africa
    1918. *Report of the Natives Land Committee, Western Transvaal.* Cape Town : Cape Times. U.G.23/1918.
    1932. *Report of the Native Economic Commission,* 1930–2. Pretoria : Government Printer. U.G.22/1932.
    1944. *Report of the Witwatersrand Mine Natives' Wages Commission.* Pretoria : Government Printer. U.G.21/1944.
    1946. *The Native Reserves and their place in the economy of the Union of South Africa.* Social and Economic Planning Council, Report No. 9. Pretoria : Government Printer. U.G.32/1946.
    1948. *Report of the Native Laws Commission,* 1946–8. Pretoria : Government Printer. U.G.28/1948.
    1949. *Population census, 1946. Vol. I : Geographical Distribution of the population of the Union of South Africa.* Pretoria : Government Printer. U.G.51/1949.
    1950. *Report on agricultural and pastoral production,* 1947–8. Agricultural census, No. 22. Pretoria : Government Printer. U.G.30/1950.
    1951. (a) *Official Year Book of the Union, and of Basutoland, Bechuanaland Protectorate, and Swaziland.* No. 25, 1949. Pretoria : Government Printer.
    (b) *Report of the Commission on Native Education,* 1949–51. Pretoria : Government Printer. U.G.53/1951.

van der Merwe, W. J.
    1936. *The development of missionary attitudes in the Dutch Reformed Church in South Africa.* Cape Town : Nasionale Pers. (Kgatla.)

van Warmelo, N. J.
    1931. *Kinship terminology of the South African Bantu.* Pretoria : Government Printer, Ethnological Publications, No. 2.
    1935. *A preliminary survey of the Bantu tribes of South Africa.* Pretoria : Government Printer, Ethnological Publications, No. 5. (Lists of tribes and their distribution.)
    1937. Grouping and ethnic history. In Schapera (ed.), 1937, pp. 43–66.

van Warmelo, N. J.—*continued.*
    1944.   (*a*) The Bakgatla ba ga Mosêtlha. Pretoria: Government Printer, *Ethnological Publications*, Nos. 17–22, pp. 3–11.
           (*b*) The BaHwaduba. Pretoria: Government Printer, *Ethnological Publications*, Nos. 17–22, pp. 23–32.
           (*c*) The tribes of the Vryburg district. Pretoria: Government Printer, *Ethnological Publications*, Nos. 17–22, pp. 33–43. (Tlhaping, Rolong, Tlharo.)

Walker, E. A.
    1940.   *A history of South Africa.* Revised edition. London: Longmans.

Wallis, J. P. R. (ed.)
    1945.   *The Matabele Journals of Robert Moffat*, 1829–60. 2 vols. London: Chatto & Windus. Central African Archives, Oppenheimer Series, No. 1. (Important source book for history of Ngwaketse, Kwena, Ngwato.)

Wande, F. O. A.
    1949.   Agro-economic survey of the Barolong Farms. Unpublished report to B.P. Government.

Wedgwood, Camilla, and Schapera, I.
    1930.   String figures from Bechuanaland. *Bantu Studies*, **4**, pp. 251–68. (Kgatla.)

Whitfield, G. M. B.
    1948.   *South African Native Law.* Second edition. Cape Town: Juta. (Includes previously unpublished material on the Hurutshe by G. P. Lestrade.)

Willoughby, W. C.
    1905.   Notes on the totemism of the Becwana. *J.R. Anthrop. Inst.*, **35**, pp. 295–314.
    1909.   Notes on the initiation ceremonies of the Becwana. *J.R. Anthrop Inst.*, **39**, pp. 228–45.
    1923.   *Race problems in the new Africa.* Oxford: Clarendon Press. (Tswana, pp. 46–138.)
    1928.   *The soul of the Bantu.* London: Student Christian Movement. (Study of Bantu religion, with many first-hand observations on the Tswana.)
    1932.   *Nature-worship and taboo.* Hartford, Conn.: Hartford Seminary Press. (Continuation of preceding work, again with original observations about the Tswana.)

LANGUAGE AND LITERATURE

Brown, J.
    1931.   *Secwana Dictionary.* (Revised, enlarged, and rearranged by J. Tom Brown.) Tiger Kloof: London Missionary Society.

Cole, D. T.
    1949.   Notes on the phonological relationships of Tswana vowels. *African Studies*, **8**, pp. 109–31.

Crisp, W.
    1905.   *Notes towards a Secoana Grammar.* (4th edition.) London: S.P.C.K.

Doke, C. M.
    1937.   Language. In *The Bantu-speaking tribes of South Africa* (ed. I. Schapera. London: Routledge), pp. 309–31. (General characteristics of S.A. Bantu languages.)
    1945.   *Bantu: modern grammatical, phonetical, and lexicographical studies since 1860.* London: International African Institute. (Bibliographical history of research in Tswana, pp. 92–3.)

Guthrie, M.
    1948.   *The classification of the Bantu languages.* London: International African Institute. (Tswana and other S.E. Bantu languages, pp. 67–70.)

Jones, D.
    1928.   *The Tones of Sechuana nouns.* London: International Institute of African Languages and Cultures, Memorandum No. 6.

Kgasi, M.
    1939.   *Thutô ke eng.* Alice: Lovedale Press. (Education, traditional and modern, with useful data on magic.)

Kuhn, G.
    1935.   Tshuana-Texte. *Z. EingebSpr.*, **26**, pp. 301–17.

Lestrade, G. P.
    1937.   A practical orthography for Tswana. *Bantu Studies*, **11**, pp. 137–48. (*Exposé* of present standard orthography.)
    1938.   Locative-class nouns and formatives in Sotho. *Bantu Studies*, **12**, pp. 35–62.
    1944.   *Some Kgatla animal stories* (with English translations and notes). University of Cape Town, School of African Studies: Communications, n.s., No. 11. Reprinted by Lovedale Press.

Letele, G. L.
  1945. The noun class-prefix in the Sotho group of Bantu languages. *Fort Hare Papers*, **1**, pp. 37–66.
Mogotsi, A.
  1931–2. Tiragalo ea morafe oa Bakoena. *Mosupa-Tsela*, vols. **18**, nos. 4, 5, 6, 11, 12 ; **19**, no. 3, etc. (History of the Kwena tribes in the Transvaal.)
Moloto, D. P.
  1943. *Mokwena*. Bloemfontein : Nasionale Pers. (Novel of tribal life.)
Norton, W. A.
  1922. Sesuto and Sechwana praises. *Trans. roy. Soc. S. Afr.*, **10**, pp. 253–66. (Texts and translations.)
Plaatje, S. T.
  1916. *Sechuana proverbs with literal translations and their European equivalents.* London : Kegan Paul.
  (1930). *Diphosho-phosho.* Morija : Printing Works. (Translation of Shakespeare's *Comedy of Errors*.)
  1937. *Dintshontsho tsa bo-Juliuse Kesara.* Revised and edited by G. P. Lestrade. Johannesburg : University of the Witwatersrand Press. (Translation of Shakespeare's *Julius Caesar*.)
Raditladi, L. D.
  1945. *Motswasele II.* Johannesburg : University of the Witwatersrand Press. (Historical drama.)
Schapera, I.
  1932. Kxatla riddles and their significance. *Bantu Studies*, **6**, pp. 215–31. (Texts, with translations and commentary.)
Schapera, I. (ed.)
  1938. *Mekgwa le melaô ya BaTswana.* Alice : Lovedale Press. (Collection of texts, by Native informants, on laws and customs ; Kgatla, Ngwato, Ngwaketse, and Kwena.)
  1940. *Ditirafalô tsa merafe ya BaTswana ba lefatshe la Tshireletsô.* Alice : Lovedale Press. (Vernacular histories of the Rolong, by Z. K. Matthews ; Kwena, Ngwaketse, and Kgatla, by I. Schapera ; Ngwato and Tawana, by G. E. Nettelton ; Malete and Tlôkwa, by V. Ellenberger.)
Seboni, M. O. M.
  1947. *Rammônê wa Kgalagadi.* Bloemfontein : Nasionale Pers. (Novel about the life of a Kgalagadi boy.)
Tucker, A. N.
  1929. *The comparative phonetics of the Suto-Chuana group of Bantu languages.* London : Longmans.
van der Merwe, D. F.
  1941. Hurutshe poems. *Bantu Studies*, **15**, pp. 307–37. (Texts, with translation and commentary.)
van der Merwe, D. F., and Schapera, I.
  1943. *A comparative study of Kgalagadi, Kwena, and other Sotho dialects.* University of Cape Town, School of African Studies : Communications, n.s., No. 9.
Wookey, A. J.
  1922. *Secwana Grammar.* (2nd edition, revised, rearranged, and enlarged by J. Tom Brown.) London Missionary Society.
  1929. *Diñwaô leha e le dipolèlô kaga dicò tsa Secwana.* Third edition. Tiger Kloof : London Missionary Society. (Histories of Tswana tribes.)

SUPPLEMENTARY BIBLIOGRAPHY
1953–1967

Middleton, Coral
 1965. *Bechuanaland: a Bibliography*. University of Cape Town: School of Librarianship (Bibl. Series).

Mohome, Paulus, and Webster, John B.
 1966. *A Bibliography on Bechuanaland*. New York: Syracuse University (Progam of E. Afr. Stud., Occ. Bibl.).

Niemandt, J. J.
 1960. Bibliografie van die Bantoetale in die Unie van Suid-Afrika. 3. Tswana. *Bantoe-Onderwysblad*, Mar.

GENERAL

Ballantine, Christopher
 1965. The polyrhythmic foundation of Tswana pipe melody. *Afr. Music*, **3**, 4, pp. 52–67.

Breutz, P.-L.
 1953. (a) *The Tribes of Rustenburg and Pilansberg Districts*. Pretoria: Department of Native Affairs (Ethnol. Publ. no. 28).
    (b) *The Tribes of Marico District*. Pretoria: Department of Native Affairs (Ethnol. Publ. no. 30). (Mainly Hurutshe.)
 1954. *Die stamme van die distrik Ventersdorp*. Pretoria: Department of Native Affairs (Ethnol. Publ. no. 31). (Fokeng.)
 1956. *The Tribes of Mafeking District*. Pretoria: Department of Native Affairs (Ethnol. Publ. no. 32). (Rolong.)
 1958. Tswana tribal governments today. *Sociologus*, n.f. **8**, 2, pp. 140–54.
 1959. *The Tribes of Vryburg District*. Pretoria: Department of Native Affairs (Ethnol. Publ. no. 46).

de Jager, E. J., and Seboni, M. O. M.
 1964. Bone-divination amongst the Kwena of the Molepolole district, Bechuanaland Protectorate. *Afr. u. Übersee*, **48**, 1, pp. 2–16.

Erasmus, D. P.
 1959. Die ekonomiese belangrikheid van trekarbeid vir Betsjoenalandprotektoraat. *S. Afr. J. Econ.*, **27**, 1, Mar. pp. 35–46.

Feddema, J. P.
 1966. Tswana ritual concerning rain. *Afr. Stud.*, **25**, 4, pp. 181–95.

Meij, L. R.
 1966. The Clark dolls test as a measure of children's racial attitudes: a South African study [Tswana]. *J. Soc. Res.*, **15**, 2 ,Dec., pp. 25–40.

Molema, S. M.
 1966. *Montshiwa, Barolong Chief and Patriot 1815–1896*. Cape Town: Struik.

Pauw, B. A.
 1960. (a) *Religion in a Tswana Chiefdom*. London: Oxford University Press for International African Institute.
    (b) Some changes in the social structure of the Tlhaping of the Taung Reserve. *Afr. Stud.*, **19**, 2, pp. 49–76.

Schapera, I.
 1955 (2nd ed.). *A Handbook of Tswana Law and Custom*. London: Oxford University Press for International African Institute.
 1957. (a) Marriage of near kin among the Tswana. *Africa*, **27**, 2, Apr., pp. 139–59.
    (b) The sources of law in Tswana tribal courts: legislation and precedent. *J. Afr. Law*, **1**, 3, autumn, pp. 150–62.
 1958. Christianity and the Tswana (Henry Myers Lecture). *J. Roy. Anthrop. Inst.*, **83**, 1, Jan.–June, pp. 1–9.
 1963. (a) Kinship and politics in Tswana history. *J. Roy. Anthrop. Inst.*, **93**, 2, July–Dec., pp. 159–73.
    (b) Agnatic marriage in Tswana royal families. In *Studies in Kinship and Marriage*, ed. I. Schapera. pp. 103–13.
 1965. Contract in Tswana case law. *J. Afr. Law*, **9**, 3, autumn, pp. 142–53.
 1966. (a) Tswana legal maxims. *Africa*, **36**, 2, April, pp. 121–34.
    (b) Tswana chiefs as innovators. *Kroniek van Afr.*, **6**, 2, June, pp. 157–68.

Sillery, A.
 1964. *Sechele: The Story of an African Chief*. Oxford: G. Ronald.

Smith, Edwin, W.
 1956. Sebetwane and the Makololo. *Afr. Stud.*, **15**, 2, pp. 49–74.

[Tracey, Hugh]
 1959. Recording tour of the Tswana tribe, October–November 1959. *Afr. Music*, **2**, 2, pp. 62–68.

van Niekerk, B. J.
 1966. Notes on the administration of justice among the Kwena. *Afr. Stud.*, **25**, 1, pp. 37–45.

van Warmelo, N. J.
    1953. *Die Tlokwa en Birwa van Noord Transvaal.* Pretoria: Department of Native Affairs (Ethnol. Publ. no. 29.)

van Zyl, H. J.
    1958. *Die Bakgatla van Mosetlha: 'n volkekundige Studie van 'n Bantoestam se ekonomiese lewe en posisie.* Johannesburg: Voortrekkerpers.

Walton, James
    1956. Early Bafokeng settlement in South Africa. *Afr. Stud.,* **15**, 1, pp. 37–43.

LANGUAGE AND LITERATURE

Cole, Desmond T.
    1955. *An Introduction to Tswana Grammar.* Capt Town: Longmans.

Cole, Desmond T., and Mokaila, Dingaan M.
    1962. *A Course in Tswana.* Johannesburg: the authors.

Mitchison, Naomi
    1966. *Return to the Fairy Hill.* London: Heinemann.
    [Describes life among the Bakgatla of Bechuanaland.]

Sandilands, A.
    1953. *Introduction to Tswana.* Tiger Kloof, Cape: London Missionary Society Press.
    1958. The ancestor of Tswana grammars. *Afr. Stud.,* **17**, 4, pp. 192-7.

Schapera, I. (tr.) and (ed.)
    1965. *Praise-poems of Tswana Chiefs.* Oxford: Clarendon Press (Oxford Lib. Afr. Lit.).

Seboni, M. O. M.
    n.d. *Kgosi Isang Pilane.* Johannesburg: Afrikaanse Pers.
    [Biography of a Kgatla chief.]
    1956. *Kgosi Sebele II.* Pretoria: J. L. van Schaik.
    [Biography of a Kwena chief.]
    1962. *Diane le Maele a Setswana.* Alice C. P.: The Lovedale Press.
    [Extensive collection of proverbs, with brief illustrations of application.]

Sotho Language Committee
    1962. *Tswana: Terminology and Orthography,* no. 2. Pretoria: Government Printer for Department of Bantu Education.

South Africa. Department of Native Affairs
    1957. *Sotho (N. Sotho, S. Sotho, Tswana): Terminology and Orthography,* no. 1. Pretoria: Government Printer.

SELECT SUPPLEMENTARY BIBLIOGRAPHY
1968–1975

Best, Alan
    1970. General trading in Botswana, 1890–1968. *Economic Geography*, **46**, 4, pp. 598–611.

Bond, C. A.
    1974. *Women's involvement in agriculture in Botswana*. Gaborone: Government Printer.

Botswana Government
    1969– *National bibliography of Botswana*. Gaborone: National Library Service.
    1972. *Report on the population census*. Gaborone: Government Printer.
    1973. *Guide to the villages of Botswana*. Gaborone: Government Printer.

Breutz, P. L.
    1967. Über den Grad der kuturellen Eigenständigkeit der Tswana unter heutigen Lebensverhältnissen. *Paideuma*, 13, pp. 11–22.
    1968. *The tribes of the districts of Taung and Herbert*. Pretoria: Dept. of Bantu Admin., Ethnol. Pubs., 68.
    1969. Sotho-Tswana celestial concepts. In *Ethnological and linguistic studies in honour of N. J. van Warmelo*. Pretoria: Dept. of Bantu Admin. Ethnol. Pubs., 52.

Campbell, A. C.
    1968. Some notes on Ngwaketse divination, *Botswana notes and records*, 1, pp. 9–14.
    1970. Bangwaketse marriage and dissolution of marriage. *Comp & internat. law J. Sthn. Afr.*, 1, pp. 212–214.
    1972. 100 Tswana proverbs. *Botswana notes and records*, 4, pp. 121–32.

Campbell, A. C., Roberts, S. A., & Walker, J. M.
    1971. *A restatement of the Malete law of family relations, land, and succession to property*, Gaborone: Government Printer.

Chambers, R., & Feldman, D.
    1972. *Report on rural development*. Gaborone: Ministry of Finance and Development Planning.
    1973. *National policy for rural development: the government's decisions on the report on rural development*. Gaborone: Government Printer.

Coertze, R. D.
    1971. *Die familie-, erf-, en opvolkingsreg van die BaFokeng van Rustenburg: met verwysing na prosesregtilike aspekte*. Pretoria: Sabra.

Cole, Desmond T.
    1968. Frédoux's sketch of Tswana grammar. *Afr. Stud.* **30**, 3–4, pp. 191–211.
    1975. *An introduction to Tswana grammar* (2nd ed.). Cape Town: Longmans.

Comaroff, Jean
    1974. *Barolong cosmology: a study of religious pluralism in a Tswana town*. Unpublished Ph. D. thesis, University of London.

Comaroff, John L.
    1973. (a) *Competition for office and political processes among the Barolong Boo Ratshidi of the South Africa-Botswana borderland*. Unpublished Ph.D. thesis, University of London.
    (b) (ed.) *The boer war diary of Sol. T. Plaatje, an African at Mafeking*. London & Johannesburg: Macmillan.
    1974. Chiefship in a South African homeland: a case study of the Tshidi chiefdom of Bophuthatswana. *J. Sthn. Afr. Stud.*, **1**, 1, pp. 36–51.
    1975. Talking politics: oratory and authority in a Tswana chiefdom. In Maurice Bloch (ed.), *Political language and oratory in traditional society*. London: Academic Press.

Curtis, Donald
    1972. The social organization of ploughing. *Botswana notes and records*, 4, pp. 67–80.

Dachs, Antony
    1968. *Missionary imperialism in Bechuanaland, 1813–1896*. Unpublished Ph.D. thesis, University of Cambridge.
    1972. Missionary imperialism—the case of Bechuanaland. *J. Afr. Hist.*, **13**, 4, pp. 647–58.
    1975. (ed.) *Papers of John Mackenzie*, Johannesburg: Witwatersrand University Press.

Feddema, J. P.
    1972. *Transvaal-Botswana kiezen voor de moderne wereld. Een vergelijkende studie van die socio-culturele verandering in een zgn. Bantoestan als antwoord op het sanbod van de 'boeren', de zending en die industriele stad.* Baarn: Wereldvenster.

Gardner, Roy
    1974. Some sociological and physiological factors affecting the growth of Serowe. *Botswana notes and records*, 6, pp. 77–88.

Gillett, Simon
    1973. The survival of chieftaincy in Botswana. *Afr. Aff.*, **72**, 287, pp. 179–85.

Johnston, Thomas F.
    1973. Aspects of Tswana music. *Anthropos*, 68, pp. 889–96.

Kuper, Adam
    1969. (a) The work of customary courts: some facts and speculations, *Afr. Stud.*, **28**, 1, pp. 37–48.
        (b) The kinship factor in Ngologa politics, *Cah. Et. Afr.* **9**, 2(34), pp. 290–305.
    1970. (a) The Kgalagadi in the nineteenth century, *Botswana notes and records*, 2, pp. 45–51.
        (b) The Kgalagari and the jural consequences of marriage. *Man*, **5**, 3, pp. 466–81.
        (c) *Kalahari Village Politics: an African democracy.* London: Cambridge University Press.
    1971. The Kgalagari *lekgota*. In Audrey Richards and Adam Kuper (eds.), *Councils in Action.* London: Cambridge University Press. (Cambridge Paps. in soc. Anthrop., 6.)
    1975. (a) The social structure of the Sotho-speaking peoples of Southern Africa. *Africa*, **45**, 1, pp. 67–81; 2, pp. 139–49.
        (b) Preferential marriage and polygyny among the Tswana. In Meyer Fortes and Sheila Patterson (eds.), *Studies in African social anthropology.* London: Academic Press.

Legassick, Martin
    1969. The Sotho-Tswana peoples before 1800. In Leonard Thompson (ed.) *African societies in southern Africa.* London: Heinemann.

Maggs, T.
    1972. Bilobial dwellings: a persistent feature of southern Tswana settlements. *Goodwin Series no. 1.* South African Archaeological Society.
    1976. Iron Age patterns and Sotho history on the southern Highveld: South Africa. *World Archaeology*, **7**, 3, pp. 318–32.

Parsons, Q. N.
    1971. The 'image' of Khama the Great—1868 to 1970. *Botswana notes and records*, 3, pp. 41–58.
    1972. *The world of Khama.* Lusaka: NECZAM/Hist. Assoc. Zambia.
    1973. (a). *Khama III, the Bamangwato and the British, with special reference to 1895–1923.* Unpublished Ph.D. thesis, University of Edinburgh.
        (b) On the origins of the Bamangwato. *Botswana notes and records*, 5, pp. 82–103.
    1974. The economic history of Khama's country in Southern Africa. *Afr. Soc. Res.*, 18, pp. 643–75.

Proctor, J. H.
    1968. The house of chiefs and the political development of Botswana. *J. mod. Afr. St.*, **6**, 1, pp. 59–79.

Reyneke, J. L.
    1972. Towery by die Bakgatla-ba Kgafela. In J. F. Eloff and R. D. Coertze. *Etnografiese Studies in Suidelike Afrika.* Pretoria: Van Schaik.

Roberts, S.
    1970. (a) Kgatla law and social change. *Botswana notes and records*, 2, pp. 56–63.
        (b) *A restatement of the Kgatla law of domestic relations.* Gaborone: Government Printer.
        (c) *A restatement of the Kgatla law relating to land and natural resources.* Gaborone: Government Printer.
        (d) *A restatement of the Kgatla law of succession to property.* Gaborone: Government Printer.
    1971. The settlement of family disputes in the Kgatla customary courts: some new approaches. *J. Afr. Law,* 15, pp. 60–76.
    1972. *Tswana family law.* London: Sweet & Maxwell. (Restatement of African Law, 5.)
    1973. Mmetlhong's Field. *Rural Africana,* 22, pp. 1–13.

Schapera, I.
    1969. (a) Some aspects of Kgatla magic. In *Ethnological and linguistic studies in honour of N. J. van Warmelo*. Pretoria: Government Printer.
    (b) Contract in Tswana law. In Max Gluckman (ed.) *Ideas and procedures in African customary law*. Oxford University Press for the International African Institute.
    1970. (a) *Tribal innovators: Tswana chiefs and social change 1795–1940*. London: Athlone Press. (L.S.E. Monogr. soc. Anthrop., 43.)
    (b) The crime of sorcery. *Proc. R.A.I.*, 1969, pp. 15–23.
    (c) The early history of the Khurutshe. *Botswana notes and records*, 2, pp. 6–13.
    1971. *Rainmaking rites of Tswana tribes*. Leiden: Afrika-Studiecentrum, Cambridge: Afr. Stud. Centre (Afr. Res. Docum., 3.)
    1974. (ed.) *David Livingstone South African papers 1849–1853*. Cape Town: Van Riebeeck Soc. (2nd ser., 5.)

Schapera, I. & Roberts, S.
    1975. Rampedi revisited: another look at a Kgatla ward. *Africa*, **45**, 3, pp. 258–79.

Sillery, Antony
    1974. *Botswana: a short political history*. London: Methuen.

Teichler, G. H.
    1971. Notes on the Botswana pharmacopoeia. *Botswana notes and records*, 3, pp. 8–11.

Tlou, T.
    1970. Khama: great reformer and innovator. *Botswana notes and records*, 3, pp. 98–105.
    1972. *A political history of north-western Botswana to 1906*. Unpublished Ph.D. thesis, University of Wisconsin, Madison.
    1974. The nature of Batswana states: towards a theory of Batswana traditional government—the Batawana case. *Botswana notes and records*, 6, pp. 57–75.

Van Niekerk, B. J.
    1968. Notes on the Kwena law of marriage, *Comp. & internat. law J. Sthn. Afr.*, 1, pp. 100–107.

Van Zyl, H. J.
    1972. Die landbou en veeteelt by die Bakgatla van Mosetlha. In J. F. Eloff and R. D. Coertze (eds.) *Etnografiese Studie in Suidelike Afrika*. Pretoria: Van Schaik.

Werbner, R. P.
    1970. Land and chiefship in the Tati concession. *Botswana notes and records*, 2, pp. 6–13.
    1971. Local adaptation and the transformation of an imperial concession in north-eastern Botswana. *Africa*, **41**, 1, pp. 32–41.

Wilson, Monica
    1969. The Sotho, Venda, and Tsonga. In *The Oxford History of South Africa*, eds. Monica Wilson and Leonard Thompson. Oxford: Clarendon Press.

INDEX

Administration : European, 49 ff. ; tribal, 48, 51 ff. ; local, 46, 53 f. Cf. assemblies, chief, committees, councils, courts, District Commissioner, family-group, finances, Government, governors, headmen, land tenure, laws, leadership, levies, Native Commissioner, Native Councils, regiments, sections, taxes, territorial organization, Treasuries, villages, ward
Adultery, 31, 42, 55
Advisers, chief's, 52 f.
Affines, relations between, 45 f. Cf. betrothal, marriage
Age composition of population, 14
Age : social distinctions, 38, 39. Cf. children, clothing, family, labour (division of), regiments, siblings
Age-sets. *See* Regiments
Agriculture, 20, 21 f., 28, 31, 40, 60. Cf. crops, exports, fields, imports, labour, land, ploughs, rainmaking, taboos, tribute
Ancestors, cult of, 45, 59 f., 61, 64, 65. Cf. family, kinship
Animal husbandry, 20, 21, 22 ff., 28, 32. Cf. cattle, cattle-posts, dairy industries, exports, goats, herding, herds, land tenure, livestock, pigs, serfs, sheep, water supplies
Army, tribal, 29, 47, 52. Cf. regiments
Assemblies, tribal, 52, 53, 54. Cf. council-place, councils, courts
Associations, modern, 48. Cf. churches, committees
Asylum, law of, 55

Banishment, 21, 35. Cf. tribe (membership)
Barter, 29. Cf. trade
Betrothal, 40, 41. Cf. marriage, uncle
Bogadi (bridewealth), 23, 41 f., 42, 48, 55
Boreholes, 24, 27
Burial, 59, 61
Bushmen. *See* Sarwa

Cattle : census, 23 ; individual holdings, 23, 30 ; uses, 23 ; herding, 24 ; kraals, 24, 47, 59. Cf. animal husbandry, *bogadi*, dairy industries, exports, fines, herds, inheritance, *mafisa*, milk, sacrifice, trade, transport, tribute
Cattle-posts, 21, 23 f. Cf. herding, land tenure, magic, servants, water supplies
Ceremonies, tribal, 22, 37, 60, 61, 62. Cf. ancestors, disease, initiation, magicians, rainmaking, religion
Chief : succession to office, 36, 37, 51 ; wives, 39, 51 ; death and burial, 60 ; retainers, 28 ; advisers, 52 f. ; duties and activities, 22, 24, 38, 51 f., 60, 62 ; powers and prerogatives, 21, 29, 46, 52, 55 ; wealth and sources of income, 22, 23, 26, 29, 31 ; economic obligations, 31, 51 ; legislative functions, 25, 29, 51, 52, 53 ; offences against, 52 ; checks on autocracy, 52. Cf. Administration, banishment, ceremonies, Christianity, committees, council-place, councils, courts, finances, Government, history, land tenure, laws, levies, magic, regiments, serfs, sorcery, tribute

Children, 37, 38, 40, 42, 60. Cf. age, betrothal, burial, descent, education, family, guardianship, household, initiation, widows
Christianity : local history, 15 f., 58 ; extent, 47 f. ; adherents, 38, 58 ; local beliefs, 58 ; effects on tribal life, 22, 39, 41, 42, 47 f., 52, 58, 61. Cf. ancestors, *bogadi*, burial, ceremonies, chief, clothing, death, divorce, education, ghosts, initiation, literature, magic, magicians, marriage, missionaries, *Modimo*, polygamy, religion, widows, women
Christians, standards of conduct, 25, 48, 58. Cf. *bogadi*, clothing, initiation, liquor, marriage, religion
Churches, 47 f., 58
Circumcision, 38, 48
Civil wrongs, 55, 56
Classes, social, 36 f. Cf. Christians, commoners, education, immigrants, leadership, literacy, marriage, nobles, serfs, teachers
Clothing, 16, 25, 27, 43, 58
Committees, tribal, 31, 32, 50, 51, 53
Commoners, 36 f., 41, 46, 47, 53, 54
Council-place (*kgotla*), 37, 47, 52, 53, 56, 60. Cf. assemblies, chief, courts, headmen, villages, ward
Councils : domestic, 40, 45, 54 (cf. family-group, ward) ; tribal, 36, 52 f., 54 (cf. assemblies) ; inter-tribal, 36, 52 f., 54 (cf. Native Councils)
Courts : chief's, 51, 52, 53, 66 ; domestic, 40, 53 ; Government, 50, 51 ; local, 46, 47, 53, 54 ; regimental, 39. Cf. trials
Cousins, 41, 45. Cf. Kinship
Crimes, 55, 56. Cf. banishment, chief, fines, law, punishments, sorcery
Crops, 21 ; trade in, 29. Cf. agriculture, exports, food, granaries, imports, magic
Cultural diversity, 34, 58. Cf. *bogadi*, ceremonies, Christianity, denominations, education, immigrants, initiation, laws (tribal), marriage, polygamy, rainmaking, regiments, sections, serfs, territorial organization, villages, ward

Dairy industries, 24, 29
Dams, 21, 24
Death : associated beliefs and practices, 59, 60, 61. Cf. ancestors
Demi-gods, 59
Demography, 11 ff.
Denominations, religious, 58. Cf. Christianity
Density of population, 11 f.
Deputy chief, 53
Descent, 40, 42, 44 f. Cf. ancestors, classes, kinship, totemism
Dialects, 17
Diet, 25. Cf. Food
Disease, treatment of, 54, 61, 62. Cf. divination, magic, sorcery
Disputes, domestic, 40, 45, 61. Cf. courts, divorce, sorcery
District Commissioner, 31, 51. Cf. Administration, Government

Districts, tribal, 36, 54
Divination, 59, 61, 62, 64 f. Cf. dreams, omens
Divorce, 42
Dogs, 22
Donkeys, 22, 23
Dreams, 59, 63, 65
Dress and decoration, 25. Cf. clothing
Dwellings, 26, 39 f. Cf. huts

Economy, 19 ff. ; changes in, 16, 22, 24. Cf. agriculture, animal husbandry, dwellings, employment, expenditure, finances, food, hunting, inheritance, labour, land, money, occupations, property, taxes, trade, transport, utensils, water supplies
Education : traditional, 38, 40 ; modern, 17, 22, 31, 32, 38, 48, 58. Cf. children, literacy, magicians, missionaries, teachers
Employment : local, 28, 30 ; abroad, 30 f. Cf. herding, labour, magicians, occupations, retainers, serfs, wages, work-party
Environment, geographical, 19 f.
European influences, impact and spread, 15 f. Cf. Administration, associations, Christianity, councils (inter-tribal), economy, education, Government, household, inheritance, labour, land, laws, liquor, marriage, religion, trade
Exchange, 23, 29. Cf. *bogadi*, chief, fields, gifts, hospitality, *mafisa*, retainers, trade
Expenditure : cash, 27 ; tribal, 31 ff. Cf. clothing, finances, fines, food, Funds, furniture, imports, levies, magicians, taxes, traders, transport
Exports, 20, 23, 29, 30. Cf. labour (migrant)

Family : stability, 31 ; property 22, 23, 24, 40 ; behaviour patterns, 38. Cf. adultery, ancestors, children, disputes, divorce, education, expenditure, household, inheritance, kinship, levirate, marriage, sorcery, sororate, widows, wives
Family-group, 40, 45, 46, 47, 53, 54. Cf. betrothal, *bogadi*, councils, courts, disputes, gifts, kinship, marriage, widows
Fields : distribution and size, 21 f. ; ownership, 21, 22, 40 ; inheritance, 42. Cf. agriculture, inheritance, land tenure
Finances, tribal, 26, 27, 30, 31 ff. Cf. chief, economy, Funds, Government, Treasuries
Fines, judicial, 27, 31, 32, 57
Food, 20, 21, 23, 24, 25 (cf. crops, gifts, hospitality) ; production of (cf. agriculture, animal husbandry, hunting)
Fowls, 22, 23
Funds, Tribal, 31, 52, 57. Cf. Treasuries
Furniture, household, 26. Cf. inheritance, utensils

Ghosts, 61
Gifts, 26. Cf. chief, exchange, kinship
Goats, 23
God. *See* Christianity, *Modimo*, religion

Government (European) : 49 ff. ; policy, 16, 22 ; powers, 20 f., 49, 52. Cf. Administration, courts, District Commissioner, economy, education, laws, Native Commissioner, Native Councils, Reserves
Government (tribal). *See* Administration, assemblies, ceremonies, chief, councils, court, finances, governors, headmen, laws, regiments, territorial organization, Treasuries
Governors, district, 36, 54
Granaries, 22
Guardianship, 43. Cf. regency, widows

Handicrafts, 27. Cf. clothing, huts, implements, ornaments, sledges, trade, utensils
Headmen : succession to office, 54 ; duties and powers, 21, 29, 46, 47, 53 f., 55 ; ranking, 37, 47, 54. Cf. districts, family-group, sections, ward
Herding, 24, 28
Herds, size of, 23, 24
History : traditional, 14 f. ; recent, 15 f. Cf. Christianity, European influences, tribe
Horses, 22, 23
Hospitality, 26, 31. Cf. kinship, wealth
Household : social composition, 39 ; dwellings, 26, 39 f. ; subsistence activities, 27 f., 40 ; property, 22, 26, 27, 29 ; organization and functions, 40 ; economic obligations, 26. Cf. children, disputes, dwellings, family, family-group, inheritance, labour, land (tenure), livestock, marriage, sorcery, utensils, wives
Hunting, 20, 21, 24 ; tribute, 26, 29, 31 ; trade in spoils, 29
Huts, 26, 27, 40, 43. Cf. cattle-posts, dwellings

Immigrants, 34 f., 36 f., 46, 47
Implements, 22, 26, 27, 29
Imports, 16, 20, 22, 26, 27, 29, 30. Cf. clothing, expenditure, food, implements, liquor, livestock, ploughs, traders, wagons
Inheritance, 38, 40, 42 f., 45. Cf. family, guardianship, kinship, land, livestock, property, serfs, succession, wives
Initiation ceremonies, 26, 38 f., 54. Cf. chief, circumcision, Christianity, regiments
Inter-tribal : councils, 36, 52 f., 54 ; trade, 29
Irrigation, 22

Joking relationship, 45
Jurisdiction, legal. *See* Courts, trials

Kgalagadi (people), 9, 14, 34, 37. Cf. serfs
Kgotla. *See* council-place
Kinship, 43 ff. ; behaviour patterns, 26, 28, 40, 45, 46, 52, 54, 56, 66 ; terminology, 38, 43 f. Cf. advisers, affines, ancestors, betrothal, *bogadi*, children, councils, courts, cousins, death, descent, disputes, divorce, family, family-group, gifts, guardianship, hospitality, household, inheritance, leadership, levirate, marriage, names, sacrifice, siblings, sorcery, sororate, succession, totemism, ward, widows, wives

INDEX

Labour: division of, 22, 24, 27 ff., 37, 38, 40; tribal, 28 f., 46, 53 f.; migrant, 13 f., 20, 26, 30 f., 48. Cf. agriculture, children, economy, family, herding, hunting, levies, occupations, regiments, retainers, serfs, work-party

Land: Native areas, 11 f., 16, 20, 49; system of tenure, 20 f., 35, 40, 51, 52, 54; shortage, 22, 24. Cf. cattle-posts, districts, fields, hunting, inheritance, natural resources, Reserves, settlements, territorial organization, tribe

Language, 17, 34

Law: European, 49; Tswana, 55. Cf. Administration, assemblies, *bogadi*, civil wrongs, courts, crimes, divorce, Government, inheritance, land, marriage, punishments, sorcery, succession, taxes, trials

Laws: tribal, 24, 25, 29, 39, 42, 51, 53, 55; Government, 29, 31, 49, 50, 51, 56, 63

Leadership, 37, 40, 48, 54, 62. Cf. chief, classes, councils, education, governors, headmen, magicians, mobility, ranking, regiments, succession, teachers

Legal status, 40. Cf. age, children, classes, law, serfs, women

Legislation. *See* chief, Government, laws

Levies, tribal, 26, 27, 31. Cf. economy, finances, taxes, tribute

Levirate, 42

Lineage, 45, 46, 54

Liquor, intoxicating, 25

Literacy, 17 f., 48; statistics, 18. Cf. education

Literature, vernacular, 17 f., 58

Livestock, 22 f.; trade in, 29; inheritance, 42 f. Cf. animal husbandry, *bogadi*, cattle, dairy industries, dogs, donkeys, exports, food, fowls, goats, herding, herds, horses, inheritance, *mafisa*, pigs, sheep, trade, transport

Locations. *See* Reserves

Mafisa (loan-cattle), 23, 28, 31

Magic, 61 ff.; varieties of ritual, 22, 24, 61 f., 65; morphology, 63 f.; political aspects, 52, 54. Cf. ceremonies, divination, rainmaking, sorcery

Magicians, 28, 37, 48, 61, 62 ff., 65. Cf. chief

Malnutrition, 25. Cf. diet, food

Marriage: regulations, 35, 37, 39, 41, 46; ceremonies, 40, 41 f., 45. Cf. adultery, affines, betrothal, *bogadi*, Christianity, divorce, kinship, laws, levirate, polygamy, serfs, sororate, widows, wives

Masculinity, ratio, 13

Meals, 25. Cf. food, hospitality

Meat, 23, 24, 25

"Medicines" (*ditlhare*), 61 f., 63. Cf. magic, magicians

Migrant labour. *See* Labour, migrant

Migrations, tribal, 14 f., 16, 36

Milk, 23, 25

Missionaries, Christian, 15, 16, 17, 21, 58. Cf. Christianity, European influences

Mobility, social, 37. Cf. classes, commoners, education, immigrants, political office

Modimo (God), 59, 61, 63, 65

Money, 20. Cf. expenditure, finances, fines, traders

Names: collective, 9; tribal, 34; personal, 45

Narcotics, 25

Native Commissioner, 33, 50 f. Cf. Administration, Government

Native Councils, 27, 49 f. Cf. Administration, assemblies, councils, Government

Natural resources, 19 f., 21. Cf. food, hunting, water supplies

Newspapers, vernacular, 18

Nobles, 36 f., 41, 46, 47, 52, 54. Cf. chief, classes

Occupations, 27 f., 30, 32, 38, 48, 62. Cf. Administration, children, employment, herding, labour, magicians, policemen, political office, retainers, serfs, teachers, traders, women

Omens, 64 f.

Ornaments, 25

Ownership. *See* inheritance, property, wealth

Pigs, 23

Ploughs, 21, 22, 23

Poisons, use of, 65

Policemen, 51, 53

Political office, 37, 54. Cf. advisers, chief, governor, headmen, leadership, succession

Polygamy, 39, 40, 43, 48. Cf. marriage, wives

Population statistics, 11 ff.

Property, 22, 23, 40, 42. Cf. clothing, crops, dwellings, fields, gifts, hospitality, inheritance, kinship, land, *mafisa*, natural resources, trade, tribute, utensils, wives

Punishments, 55, 56, 57, 66. Cf. banishment, courts, fines, sorcery, trials

Rainmaking, 22, 60. Cf. chief, magicians

Ranking, system of, 36 f. Cf. age, classes, leadership, mobility, sexes

Regency, 51

Regiments (age-sets): creation, 38; names, 39; organization, 39, 46, 52; activities, 24, 28 f. Cf. age, army, initiation, labour (tribal)

Religion, 37, 40, 45; traditional cults, 59 f. (cf. ancestors, ceremonies, chief, *Modimo*); modern beliefs and practices, 58, 61 (cf. Christianity, churches)

Reserves, tribal, 11, 16, 20 f. Cf. land, tribe

Residence, law of. *See* banishment, divorce, family-group, marriage, sorcery, villages, ward

Retainers, hereditary, 28, 46. Cf. chief, commoners, ward

Revenues, tribal, 31, 32. Cf. economy, finances, Funds, levies, taxes, Treasuries, tribute

Sacrifice, 59, 60, 61. Cf. ancestors, ceremonies, magicians

Sarwa (Bushmen), 9, 14, 34. Cf. serfs

Schools. *See* Education (modern)

Sections, tribal, 47, 53, 54
Segmentation. *See* family-group, section, tribe, ward
Serfs, 28, 37
Servants, 28. Cf. herding, retainers
Settlements, 35 f. Cf. villages, ward
Sexes : numerical proportions, 13 f. ; social distinctions, 37 f. Cf. assemblies, Christianity, clothing, death, education, family, household, kinship, labour, marriage, occupations, political office, succession, widows, wives
Sheep, 23. Cf. *bogadi*
Siblings, 45
Sledges, 23, 26
Social classes. *See* classes, social
Social mobility. *See* Mobility, social
Sorcery, 42, 55, 56, 60, 62, 63, 65 f. Cf. divination, magicians
Sororate, 40, 42
Specialization, labour, 27 f. Cf. occupations
Spells, magical, 64
Spooks (*dipoko*), 61
Succession, law of, 42, 45. Cf. chief, descent, family-group, headmen, household, kinship, leadership, political office, wives

Taboos, 21, 22, 25, 35, 61
Taxes, 27, 31, 32, 49 ; collection, 50, 52, 54. Cf. finances, levies, revenues, tribute, Treasuries
Teachers, 30, 37, 38, 48. Cf. education (modern)
Territorial organization, 35 f., 37, 54. Cf. family-group, land (tenure), sections, serfs, settlements, villages, ward
Tobacco, 25
Totemism, 34, 35
Trade, 29 f. Cf. economy, exchange, exports, imports
Traders, 20, 29, 31
Transport, modes, of, 23, 26, 27

Treasuries, Tribal, 30, 32 f., 52, 57. Cf finances
Trials, judicial, 47, 56 f., 66. Cf. courts, law
Tribe : definition, 10, 34 ; ethnic composition, 9, 11, 34 ; name, 34 ; origins, 15, 34 ; size, 34 ; expansion, 34 f. ; membership, 34 f., 36 ; disintegration, 15, 34 ; territory, 20 f., 34 ; administration, 30, 32 ; finances, 26, 27, 30 31, ff. Cf. Administration, chief, Government
Tribes, distribution of Tswana, 10
Tribute, 22, 26, 28, 31, 54
Tswana : name, 9 ; demographic data, 11 ff. ; distribution, 9, 11 f., 17 ; subdivisions, 9 f. ; list of tribes, 10

Uncle, maternal, 42, 43, 45, 59
Utensils, household, 22, 26, 27, 29

Villages, 21, 35 f., 46 f., 53, 54

Wages, 28, 30. Cf. employment
Wagons, 23, 26
Ward, 28, 36, 45, 46 f. ; size and membership, 46 ; origins, 46 ; settlements, 46 f. ; government, 53 f. Cf. headmen
Warfare, 29
Water supplies, 19, 21. Cf. boreholes, cattle-posts, dams, irrigation, villages, wells
Wealth, social importance of, 23. Cf. cattle, chief, gifts, hospitality, *mafisa*
Wells, 21, 24
Widows, 40, 42, 43. Cf. guardianship, levirate
Witchcraft, 65. Cf. sorcery
Wives, ranking of, 40, 51. Cf. polygamy
Women, social status, 37 f. Cf. Christianity, education, inheritance, labour (division of), political office, sexes
Work-party, 28

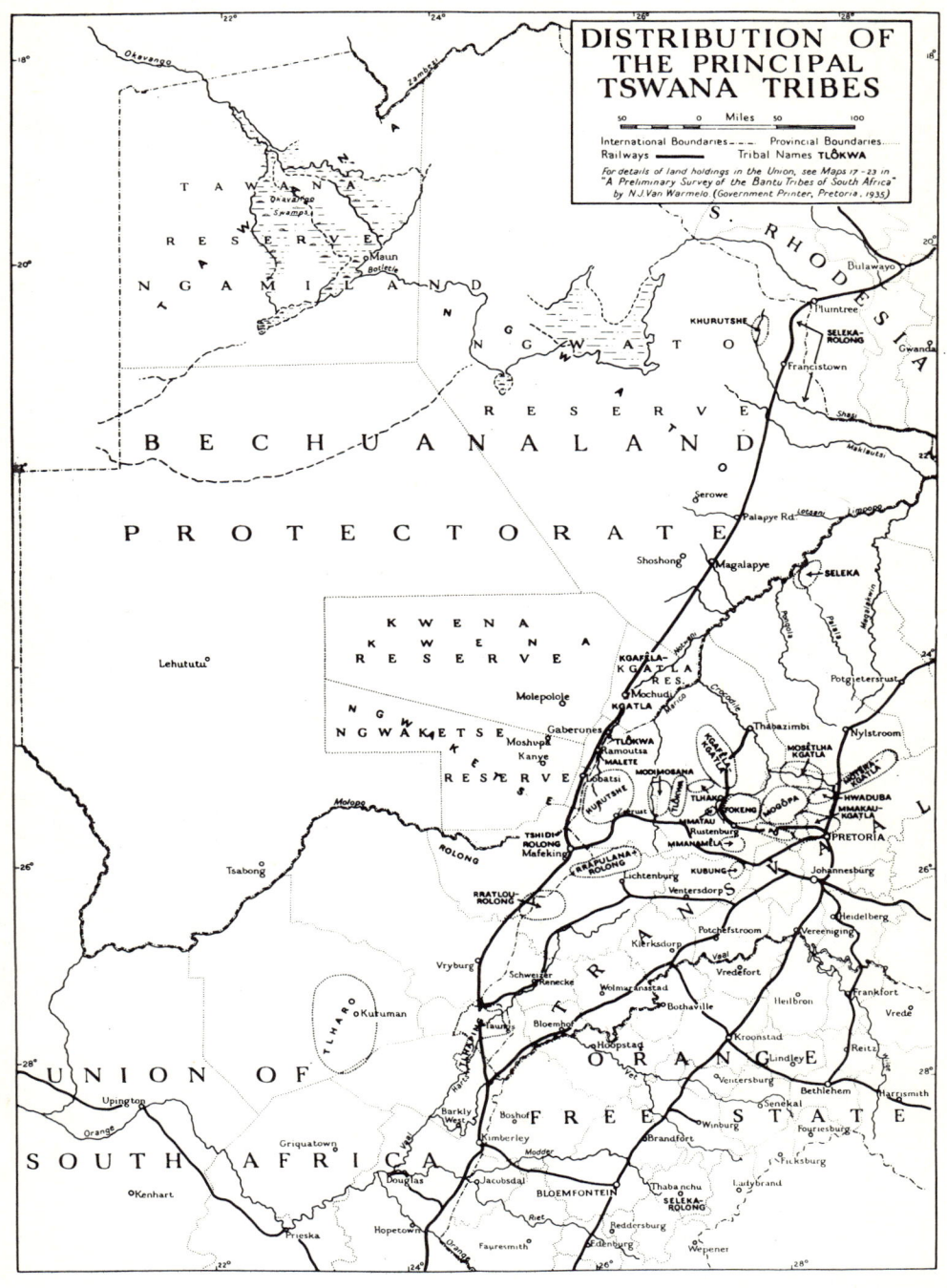